MAN'S PLAGUE?

Insects and Agriculture

by

V. G. Dethier

THE DARWIN PRESS, INC.
Princeton, New Jersey

Library of Congress Cataloging in Publication Data
Dethier, Vincent Gaston, 1915-
  Man's plague?
  Includes bibliographies and index.
  1. Insects, Injurious and beneficial 2. Agri-
culture. 3. Insect control. I. Title.
SB931.D47   632'.7   75-15216
ISBN 0-87850-026-X

This book was set in Garamond typeface
with Optima headings
Designer: Albert McGrigor

Printed in the United States of America

See, I give you every seed-bearing plant all over the earth and every tree that has seed-bearing fruit on it be your food; and to all the animals of the land, all the birds of the air, and all the living creatures that crawl on the ground, I give all the green plants for food.   Genesis 1:29-30

When I brought you into the garden land to eat its goodly fruits, you entered and defiled my land, you made my heritage loathsome.
                                               Jeremiah 2:7

Thou, Nature, art my goddess; to thy law My services are bound.
                                  Shakespeare, *King Lear*, Act I Scene 2

Contents

Illustrations

The following illustrations appear as a group after page 128.

Pl. 1. Part of a swarm of Desert Locusts (*Schistocera gregaria*) which covered 400 square miles in Ethiopia in October, 1958 (Photograph by C. Ashall).

Pl. 2. Brachonid (a parasitic wasp) about to lay eggs in a wood-boring beetle larva under the bark of a dead acacia (Photograph by Dr. Edward S. Ross).

Pl. 3. A brown lacewing *(Sympherobius angustus)* feeding on aphids (Photograph by Dr. Edward S. Ross).

Pl. 4. A ladybird beetle *(Hippodamia convergens)* feeding on aphids. At the left are some of the beetle's eggs (Photograph by Dr. Edward S. Ross).

Pl. 5. Robberfly and a grasshopper it has captured (Photograph by Dr. Edward S. Ross).

Pl. 6. Eastern Spruce Budworm moth *(Choristoneura fumiferana)* (Photograph by D. C. Anderson, Courtesy of the Canadian Forest Service).

Pl. 7. Larva of the Eastern Spruce Budworm (Courtesy of the Canadian Forest Service).

Pl. 8.   Damage to a mixed spruce fir forest by the Eastern Spruce Budworm (Photograph by D. C. Anderson, Courtesy of the Canadian Forest Service).

Pl. 9.   A modified moth scale with an elaborate evaporative surface that facilitates dispensing male sex pheromone. From a nocturnal moth *(Rusina verberata)* (Courtesy of J. Clearwater and G. Braybrook, University of Alberta, and the *Journal of Morphology*, 146[1]: 129).

Pl. 10. Rice Weevil *(Calandra oryzae)*, a pest of stored grain, magnified 150 times (Courtesy of Dr. C. Pitts).

Foreword

This book can be likened to a triangle: one side *insects*, another side *plants*, and the third side *man*. It is a good book because the author is a specialist on insects and can write well. It is an especially good book because the author knows also a great deal about plants.

Obviously he loves the animate part of our world, and hopes to influence man to use restraint in managing that part of it. He gently informs readers that lack of restraint shortens the existence of the animate part of nature, including man, on this planet.

The reader may infer that, under any political system *Homo sapiens* (Man) has yet devised, this planet's chances of return to health are decreasing annually. Witness the average male's goal of crass financial gain in a single lifetime, the average female's willingness to beget more than her quota of offspring thereby preventing zero-population growth, and our political leaders' education—learning in law courses, principally man-made laws, instead of leavening this schooling with science courses in order to learn also some laws of nature. These three circumstances are formidable obstacles to orderly evolutionary progress.

Author Dethier does not say precisely those things, but he does deplore the current degree of reliance on chemical, instead of natural, control of insects, and suggests restraints in a fashion designed to educate the uninformed. Therefore, his book would be a reasonable place for sincere political leaders (legislative, judicial, and administrative) to begin remedial reading, and a place just as useful to a concerned public.

He champions restraint in poisoning our planet. He brings light to many dark places. Nature is his goddess and to her welfare his services are bound. Long life to him and to the increasing percentage of his kind.

E. Raymond Hall
Professor of Systematics
    and Ecology (Emeritus)
University of Kansas
Lawrence, Kansas

Preface

Agriculture began as an inspired invention to feed mankind. It has become a business for the enrichment of many people who neither sow nor reap. It has given rise to a philosophy which holds that anything that obstructs the goals of agro-business must be eliminated. In this context insects have been cast in the role of archvillains, and a giant industry dedicated to their extermination has evolved. One consequence of this modern attitude toward agriculture and insects has been the impoverishment and destruction of the environment. The crusade against insects has been justified as a crusade against famine.

The purpose of this book is to inquire into the relations between insects and crop plants and between agriculture and famine. Is the insect a villain or a scapegoat? Is it so competitive with man that it must be eliminated even at the cost of destroying the environment? To what extent is agriculture geared to hunger and malnutrition?

In seeking answers to these questions we are dealing with three forms of life that dominate the earth: green plants, insects, and men. Without green plants there would be no life as we know it. When rain falls on the desert, when the volcano cools, when the glacier melts, the green plant appears. Let the highways be idle, then the green plant bursts through the concrete; the flower grows in the crannied wall; the jungle strangles the lost cities of ancient civilizations. From outer space our planet is blue, the planet of water, but that which is not blue, or desert, is green, a magical green that blankets the earth, a cover of chlorophyl that alone can efficiently capture the energy of the sun and give to the earth its oxygen, its carbon dioxide, and its food. Truly the green plant rules.

Because the earth is green it bears other forms of life. Ninety-seven percent of all animal species are insects, over a million variations on a single theme, dominant in species, in numbers, in distribution. From pole to pole, from desert to rain forest, from mountain to seashore, they live and court and reproduce. Only the open seas are barren of insects. They have become a dominant group despite the fact that they are small, weak, and short-lived. Few can profit by experience. They have no memory, no intelligence, and no culture.

Man is the third dominant form of life. He is not a noble zoological specimen. He is naked; he is vulnerable to the rigors of environment; he is weaker than a chimpanzee, lacks the visual acuity of an eagle, the finely tuned ear of a bat, the olfactory keenness of his dog. He rules with his intellect. He can will the destruction of the planet and possesses the power to wreak his will.

Man, plant, and insect! These three dominate and compete. Man and insect compete for the green plant; the green plant struggles against them both. Modern man exploits the plant as never before. He sees the insect as an enemy. In the process of trying to destroy the insect he is well on the way to destroying his environment and himself.

Full understanding of the complex interactions of plants, men, and insects can emerge only in the perspective of their antecedents. And we can plan the future only in terms of the past. This book is the story, past and present, of men, plants, and insects; the story of our present agricultural and environmental dilemma. There is a solution, but do we have the wisdom and the moral vigor to seek it?

V. G. D.
Sussex, England

Acknowledgments

The author gratefully acknowledges permission to quote from the following:

Concepts of Pest Management, by R. L. Rabb and F. E. Guthrie (eds). North Carolina State University, Raleigh, 1970. pp. 183, 117, 4, 114, 115, 164, 166, 183, 238, 115.

Bulletin of the Entomological Society of America: 1972, Zimdahl ("Pesticides—A Value Question") and Borden ("Changing Philosophy in Forest-Insect Management"). J. H. Borden, p. 269.

Biological Control of Insect Pests and Weeds, by Paul DeBach (ed.). Chapman and Hall, London, 1970. p. 712.

"Plant Pests and Diseases: Assessment of Crop Losses," by Chiarappa, Chiang, and Smith; in *Science*, 19 May 1972, pp. 769-73.

Pesticides and the Environment, brief published by the Entomological Society of Canada, Ottawa, Canada, 1970, pp. 4, 5.

Drawings adapted from *Insects, the 1952 Yearbook of Agriculture*, courtesy of the U.S. Department of Agriculture.

The research upon which this book is based was made possible by a fellowship from the John Simon Guggenheim Memorial Foundation for whose generosity I am most grateful. It is also a pleasure to acknowledge the hospitality and assistance of the University of Sussex, Brighton, England, and to offer my special thanks.

Chapter One
Origins

1
Origins

Let us imagine for a moment that we are living not in the present but two hundred million years ago, and that we have joined the British geographer C. E. P. Brooks on an imaginary voyage. We find ourselves in a genial world in which two of the three competing organisms with which we are concerned, plants and insects, have already achieved world dominance. Man has not yet arrived on the scene. The voyage begins at the equator and sets course for the North Pole. Throughout the venture the air is warm and moist, the skies sunny, the ocean calm. On the horizons the land is always low, the greatest elevations being gently rounded hills. Scenery and conditions remain essentially unchanged all the way to the Pole. The lands, more like low islands, are interlaced with broad oceanic channels.

It was during this geologic period of felicitous climate, the Mesozoic and early Tertiary, that flowering plants originated, evolved, and diversified at a measured pace, distributing themselves widely over the globe. Except for a brief (geologically speaking) cold period preceding the Eocene, the world had been one large greenhouse for millions of years. In this continuous, uninterrupted growing season, the flowering plants simply added more and more wood perennially. Their fossils reveal that they had no annual growth rings. Figs, laurels, sycamores, eucalyptus, palms, and other typically warm climate species were abundant. Everywhere forests of elms, oaks, maples, magnolias, and palms began to rival in extent the conifers which earlier still had largely replaced giant ferns, clubmosses, and cycads.

During this same period, the primitive dragonflies and cockroaches that had populated the great Carboniferous coal forests underwent a degree of evolution comparable to that of plants. The evolution of many new species was accompanied by geographical expansion. For nearly 130 million years, the flowering plants and insects developed intimate and intricate relationships for survival. We could say, figuratively, that it was a time of mutual understanding and adjustment. Neither was threatened by excessive competition from other forms of life. The great predators and herbivores that we know, the mammals and birds, had not yet appeared on the scene and the herbivorous dinosaurs were leaving the stage.

Then suddenly, as had happened in the past, the earth entered once more into a period of intense mountain building, uplifting, and volcanism. The consequent deepening and channeling of oceans and the barriers thrust into the air by rising mountain ranges altered oceanic and atmospheric circulation. These changes, together with the dust thrown into the atmosphere by volcanoes, cumulatively lowered temperatures, and the glaciation of the Quarternary epoch began, about one million years ago. Now there were great altitudinal and latitudinal differences in temperatures. The advancing glaciers reduced living space, and for the first time since their appearance on earth, flowering plants were placed under intense environmental stress. Where they could they retreated from the advancing ice and cold. Evolutionary criteria changed. Whereas the plants had for millions of years lived in equilibrium with their environment, they were now put to the test. Many species retreated to the equator, many became extinct, and woody types were replaced by herbaceous plants more suited to the fluctuating, harsh climates of the glacial period. Today, the survivors present a different picture in terms of distribution and composition of species than during the warm preglacial Mesozoic and early Tertiary. As Good remarked in *The Geography of the Flowering Plants*, "the distribution of plants today unquestionably suggests that the Flowering Plants are recovering from a catastrophe, and that they are actively in the process of reconstituting that generalised balance or equilibrium between vegetation and environment which (was) the outstanding feature of preglacial plant geography."

Changed though they were, the flowering plants survived the most severe challenge that climate could present. They had already made their peace, so to speak, with herbivorous insects. The two now settled down to renewed relationships in a world that, while still in the glacial period, was moving to a new phase within the cycle. It is at this moment in prehistory, after plants and insects have lived in equilibrium for nearly 200 million years, that the third character in our story makes his appearance.

MAN'S DEBUT

Although Man appeared approximately seven to four million years ago, it is most unlikely that he exerted any evolutionary influence on plants or insects at that time. He invented agriculture and learned to form a partnership with plants only about 10,000 years ago. Most of the first plants cultivated were "multi-purpose" plants that supplied such principal nutrients as starch and sugar. Other plants were cultivated because they were a source of fiber.

Cultivation presumably was developed independently and in a very short time, historically speaking, in several parts of the world. Plants of southeastern Asian origin included banana, turmeric, taro, some yams, sago palm, sugar cane, derris, breadfruit, citrus plants, and persimmon; all are propagated by cuttings. From Mediterranean shores came the date palm, olive, and fig —also capable of cultivation by cuttings. From the Caucasus area came the grape. A subordinate center of vegetative culture bordered the Gulf of Guinea. Here were cultivated African yams. Other African plants of importance today such as plantain, banana, and taro came from Asia, probably by way of southern Arabia. In the New World, conditions similar to those postulated for Asia occurred in the lands of Mesoamerica bordering the Caribbean on the south. Here grew manioc, sweet potato, yams, arrow root, plants that furnished narcotic drinks, tobacco, cocas, and *Lonchocarpus nicon*, the source of the insecticide rotenone, and in earlier times a fish poison. With the exception of the tobaccos and cocas these plants are all propagated by cuttings.

Thus, very early in the history of agriculture, but very late in the evolutionary development of the flowering plants, man began by vegetative culture to select plants suited to his particular needs. By this time, the world probably had a human population of only a few million, still thinly spread throughout the Old World. So thorough has been the domestication of these plants that many are quite incapable of reproducing by themselves.

The domestication of plants through the selection of seeds was a later development. It involved annual harvesting, storing, and sowing in a precise and seasonally directed sequence. Accordingly, all the domesticated seed plants are annuals or became annuals through cultivation, and the same trend toward an inability to reproduce without man's assistance developed in plants propagated by seeds.

Vegetative propagation was best suited to the warmer, moister regions of the world. Climate tended to restrict it to these areas. On the fringes of the "planter" culture, that is, culture by propagation, the newer technique of annual seed sowing and harvesting replaced it. Small valleys and adjacent slopes may have provided the first plots. Originally seed plants were weeds, not cultivated, but protected because of some desirable quality, just as wild tomatoes are in some parts of Mesoamerica today. As Sauer pointed out, when climatic advantages shifted from plants propagated vegetatively to those propagated by seeds, the gardener started to select the latter. He centered his interest on three kinds of seeds: those containing principally starch, those with protein, and those rich in oil. Starchy grains were derived from weedy grasses and protein and fat from legumes.

Three principal centers of seed domestication have been identified: North China, the Near East, and Mesoamerica. There are numerous subordinate centers or areas of diversification. From northern China came millets, the small, many-seeded grasses that were the progenitors of the basic grains. One of the millets occurring in northern China in Neolithic times (*Panicum miliaceum*) was likewise cultivated by the Swiss Lake Dwellers. China also produced the soybean. From the Ethiopian highlands came the great millet, durras, barley, sweet sorghums, some lentils, and sesame. Western India produced true peas, chick peas (garbanzo), lentils, and some wheats. The primitive wheat, einkorn, came from Asia Minor, and another kind of wheat, emmer, from Syria and Palestine. Mesoamerica produced maize, four kinds of beans—scarlet runner, kidney (navy, frijol), lima, and tepary (Indian bean of Sonora)—and squashes and pumpkins. The latter were originally cultivated for their edible seeds. Many of these plants had multiple uses. Wheat and barley, for example, provided straw for thatching and bedding. Flax and hemp provided both oil and fiber.

Wheat illustrates a particularly interesting development because it is composed of at least three basic groups. These groups are subdivided on the basis of structural properties, chemical characteristics that affect baking, and the number of chromosomes. The group that is basic to all contemporary wheats is an ancient one containing two sets (diploid) of seven chromosomes. The other groups have four (tetraploid) and six (hexaploid) sets respectively. The ancient Neolithic cereal, now almost disappeared, was the diploid *Triticum*, einkorn. From hybridization with its own weeds, possibly the quack grasses (*Apropyron*) and the weed grasses (*Aegelops*), it produced in Neolithic times emmer and the so-called macaroni wheats, sticky and protein-rich tetraploid grains. The original cross would still have been diploid, as wild emmer is, but chromosome doubling, a not uncommon occurrence in plants, would have produced the domesticated emmer, once the most common cultivated wheat. Because of its fragile heads, it is not widely cultivated today except for cattle feed. Durum, another tetraploid, does not have the seeds so tightly clasped by the chaffy bracts (glumes) as to prevent thrashing. It is grown widely in the United States as a source of macaroni flour. By Bronze Age times, other wheats had been developed. Swiss Lake Dweller wheats were probably related to Persian wheats (diploid). By the Iron Age, the hexaploid bread wheats and spelt had been developed. These are the modern wheats. The bread wheats have tough non-shatterable stalks and seeds loose in the glumes so that they are economically harvested and threshed. Spelt has tight glumes.

The hybridizations of early times did not result from human design. The basic cereals and appropriate weeds just happened to be growing intermixed. Man took advantage of the propitious crossing to select and sow the new hybrids. This is in contrast to the modern development of wheats in which plant geneticists actually manipulate the hybridization. It is interesting that spelt can be synthesized by crossing emmer and the wild *Aegelops*. None of these modern wheats exists in the wild form.

A number of grains reached cultivation as weeds tolerated in the main crop. Originally, rye grew as a weed in fields of soft wheat; reapers could not separate the two. In regions where winters were mild, wheat constituted the greatest number of plants. Where winters were severe, rye surpassed wheat. As farmers migrated northward carrying mixed seed, rye became the principal crop. Similarly, in the British Isles oats started as a weed growing in fields where emmer was cultivated, and ended up as a separate crop in the Scottish highlands.

In the early domestication of plants, the characters of the plant species that interested man were most often not the ones that were of the greatest survival value to the plant in its battle against pests, adverse climate, and poor soil conditions. This is illustrated by the following case: Neolithic man selected grains not for the size of the grain but for the number of seeds per head and for non-shattering heads. When the seed of wild grains (and plants with pods) ripens, the head shatters or the pod bursts open scattering the seed. Reaping is not possible under these circumstances. Man must work rapidly and at just the precise time in order not to lose seed. Gathering was done by shaking the stalks over baskets or, as is done in harvesting wild rice today, into boats. In ancient Egypt, it was possible to reap, after a fashion, the emmer grown there by grasping a bunch of stalks just beneath the heads and cutting farther down. In this way not too many of the fragile heads broke off.

Early agriculturalists were successful in selecting mutations in which the heads would not shatter nor pods open. Normally, these mutations could not survive without human intervention since they could not disperse seed. From the human point of view, the mutation was most valuable because the crop would stand until man chose to harvest it. He could collect by reaping at his own pace. Then, still more or less at his leisure, he could subject the grain to threshing or flailing to separate the seed from the stalk and glumes. The harvested grain could be dried, stored indefinitely, transported at will, planted whenever and wherever desired, and some kept as insurance against bad years.

In the New World, the great cereal was maize. The identity of its wild ancestor is still not known. According to the oral tradition of the Nahautl Indians, the red ant brought corn seed back from the underworld in response to an injunction by the gods to seek food for man. The most ancient maize found so far has been excavated in a cave in southern Mexico. These ears, not even one inch long, consist of fragile cobs bearing a few grains with long glumes. At the tip of the cob are the remains of male flowers, the cobs themselves being lateral growths from these. In modern corn the ears are truly lateral from the main stem, which bears terminal male flowers, the tassels. The ear consists of many female flowers in rows on a cob. Each flower produces one seed, a kernel, and a long silk which is the stigma and style.

The little wild maize grew about 7,000 years ago. By A.D. 700, only domesticated varieties existed. Probably about 1500 B.C., an unprecedented process of variation began. Extensive hybridization with the annual grass teosinte occurred. Today, wild teosinte continues to infuse new genes into cultivated maize in Mexico. It is also possible that wild perennial grasses of the genus *Tripsacum* contributed to the evolutionary development of maize; the two hybridize readily in the laboratory. At the time Columbus discovered America, in 1492, there were a great number of varieties of maize. Today, genetic manipulation—principally selection for yield and the development of hybrid corn—has reduced commercial corn to a few varieties that are nutritionally and genetically inferior to the many ancient kinds. They are specialized to grow in rich soils to feed cattle.

THE DISPERSAL OF PLANTS

One of the great difficulties in tracing the origins of domesticated plants is the fact of dispersal. Man is a restless animal, and from the prehistoric past to the present he has wittingly or otherwise carried his plants with him—even today despite the vigilance of hawk-eyed customs inspectors. As one plant geographer, Anderson, remarks in *Plants, Man and Life*, "Man takes his landscape with him." While origins represent one aspect of the background of the man-plant relationship, dispersal represents the other. Many domesticated plants are now living in parts of the world far from home. As already mentioned, cultivated bananas, plantains, arrowroot, and taro reached Africa from India in ancient times by way of southern Arabia. Sago palms, sugar cane, and many other endemic Indian plants spread into the Pacific as far eastward as Easter Island. Rice, bamboos, bananas, taro, persimmons, and yams moved into China and Japan. Coffee, a native African plant, is now grown principally in the New World. Cocoa, a New World plant, is now cultivated mainly in Malaya, Indo-China, and Africa. Peanuts, pineapples, manioc, the sweet potato, tobacco, rubber, maize, and cocoa all moved from their origin in the New World to the Old.

All sorts of grasses, vegetables, and weeds have moved from Europe to the New World. Most of the common grasses, blue grass, creeping bents, fescues, and redtops of eastern North America are European. Many were transported unintentionally, but some were brought by design. Timothy grass is one such example. In 1731, a New Hampshire farmer, Timothy Hanson, sailed to Bordeaux, France, on a wool ship from Boston. There in French pastures he noticed a handsome grass. Gathering seeds he returned home, planted them in one of his fields, and for four years harvested and replanted. At this point he gave the grass a name, Timothy, and began peddling seeds from New England to Maryland. Within his lifetime Timothy, the grass, became the major hay crop in the northern and central American colonies.

Most of the California grasslands are of Mediterranean origin, having first come with the Spaniards. As Anderson remarked, these are the same plants that the Greeks waded through at the siege of Troy and the same weeds that grow in the ruins of Carthage.

The spice trade of ancient civilizations and of the fifteenth and eighteenth centuries created a world-wide dispersal of plants despite the navigational, commercial, and political obstacles of the times. Because of the value of spices (the total cost of Magellan's trip around the world was reimbursed by the sale of spices brought home by the ship), controlling interests stopped at nothing to maintain their monopoly. Nutmeg and cloves were a Dutch monopoly in the seventeenth and eighteenth centuries until some hardy adventurers succeeded in smuggling them out of the Dutch-contolled Moluccas, known as the Spice Islands. Cloves were transplanted to Mauritius, Réunion, and Zanzibar.

Coffee is a picturesque example of plant dispersal. The plant is Abyssinian in origin. It may also have occurred wild in Arabia. But because of the close ties between Abyssinia and Yemen in the first few centuries of the Christian era, it is difficult to untangle the story fully. In any case, to the Arabians goes the credit for discovering the use of coffee as a beverage and first cultivating the plant in A.D. 575. As coffee became popular in the fifteenth and sixteenth centuries, the Arabians preserved their monopoly by preventing the export of viable berries. All were placed in boiling water or parched. Sometime in the seventeenth century a Moslem Indian returning home from a pilgrimage to Mecca succeeded in smuggling out live berries, which he planted outside of his hut in Mysore. Meantime, in 1616, the Dutch had succeeded in transplanting a plant from Mocha to Holland. In 1658, they began cultivation in Ceylon. The first coffee introduced into Java was shipped from Malabar, India, at the suggestion of Nicholaas Witsen, Burgomaster of Amsterdam. Earthquake and

flood destroyed the plantings, but a second attempt in 1699 succeeded. In 1706, a coffee plant was sent from Java to the Amsterdam Botanical Gardens. Offspring of this plant found their way
into most of the major botanical gardens and conservatories of
Europe. Another famous coffee plant was sent, after appropriate
governmental negotiations, by the Burgomaster of Amsterdam
to Louis XIV of France and installed ceremoniously in the Jardin
des Plantes, Paris. Offspring of this plant supplied the stock of
the coffees of the New World.

The idea of introducing coffee to the New World is credited
to Gabriel Mathieu de Clieu, a naval officer about to return to
duty as an infantry captain in Martinique(29). The first obstacle
to his plan lay in obtaining plants. He circumvented this hurdle
with the assistance of the royal physician, M. de Chirac, through
"the kindly offices of a lady of quality to whom de Chirac could
give no refusal." De Clieu embarked at Nantes in 1723. "It is
useless," he wrote many years later, "to recount in detail the
infinite care that I was obliged to bestow upon this delicate plant
during a long voyage, and the difficulties I had in saving it from
the hands of a man who, basely jealous of the joy I was about
to taste through being of service to my country, and being unable
to get this coffee plant away from me tore off a branch."

The vessel eluded a corsair out of Tunis, survived a violent
storm, only to become becalmed to the point of nearly running
out of water.

"Water was lacking to such an extent that for more than a month
I was obliged to share the scanty ration of it assigned to me with
this my coffee plant upon which my happiest hopes were founded
and which was the source of my delight. It needed such succor
the more in that it was extremely backward, being no larger than
the slip of a pink."

Fifty-four years later the plant had begotten 18,791,680 descendants in Martinique.

A little more than two hundred years after the Dutch introduced
coffee into Ceylon, in 1870, a fungus leaf-spot disease wiped out
the Ceylonese coffee plantations, forcing the Oriental Bank into
bankruptcy. As a result, Ceylon abandoned coffee growing forever. Tea took its place.

II

It is apparent from the foregoing section that man has been responsible for significant changes, and tremendous dislocations, in the
space of a few thousand years in the relationships that evolved between plants and insects over a period of many millions of years.
Much later, as we shall see, the fever of colonization in the seventeenth and eighteenth centuries and improved transportation accelerated this process beyond belief. The plants that we seek to
protect today are not the same plants nor are the conditions under
which they are grown the same.

SOME EFFECTS OF PLANT DOMESTICATION

Dispersal and domestication have wrought strange changes, and
a true mutualistic relationship has developed, one in which man
has become dependent upon plants and they upon him. The first
change is in the plants themselves. Originally by vegetative and
seed selection and lately by genetic manipulation, man has produced what are from an evolutionary point of view botanical
freaks. He has bred plants for genetic characters of interest to
him, generally without regard to the plant's survival on its own.
Maize has been bred for yield and for survival in particular soil
conditions selected by man. Commercial corn is genetically poor
compared to the many varieties grown by Central American
Indians, and because of genetic manipulation, it is no longer
capable of reproducing itself without man's assistance. Wheat,
and most grains, have been bred for size of head, for lack of shatter,
and for growth in unaccustomed climates. In the New World,
many wild varieties of potatoes disappeared with the Spanish
Conquest, and what is left are pale conformists bred for high
yield. Modern tetraploid cotton is so artificial that it survives only
at man's pleasure.

Many other cultivated plants can reproduce only with man's
midwifery. Consider, for example, the edible banana. This, the
most important fruit in international commerce, originated in
Asia, where it was probably cultivated as a root crop. The fruit,
crammed with seeds, could hardly have been edible. Over the
years, man developed banana plants that produced, without fertilization, the seedless fruit that we know today. It is propagated

by young plants that bud from the underground stem of the older plant. Bananas—the word is of African origin—spread across the Pacific about the time of Christ. In Medieval Europe the banana tree was known as the "Tree of Knowledge of Good and Evil in the Garden of Eden." It reached the New World at Santo Domingo about the year 1516. The progenitor of the great plantations of Central America that constitute the major source of the world's banana supply was a plant, the Gros Michel, introduced into Jamaica by Jean François Pouyat, a botanist, from his coffee plantation in Martinique.

In short, whereas plants would normally be co-evolving with their environment and with their pests to establish balances, man breeds plants that often are quite out of balance with their environment—in a word, are more susceptible to pests and disease. The situation is comparable in some respects to man's own relation to his diseases. Not many generations ago people who were susceptible to such diseases as tuberculosis, infantile paralysis, etc. died, leaving a residual population with a certain amount of resistance (immunity). This phenomenon was shockingly illustrated by the decimation of American Indians in the nineteenth century by tuberculosis, transmitted by the European colonists, and by the inroads of syphilis in Europe when it was introduced from America sometime around the fifteenth century. With the development of vaccines and antibiotics, susceptible individuals no longer die, so we are now artificially maintaining a mixed population of immunes and susceptibles. So it is with domesticated plants. Domestication has produced plants that are susceptible to disease and parasites. Man having produced them must now protect them.

This trend toward the maintenance of artificial populations is further hastened by the elimination of genetic wealth (i.e., nearby wild stock) in the areas where the various domesticated plants originated. Under normal circumstances many cultivated plants would still be able to evolve by hybridizing with their wild relatives growing nearby. Quite aside from the casual introduction of wild genes, man can introduce them by design to improve the genetics of his domesticated varieties. The gene pools in the centers of diversity, the ancestral genetic wealth, far exceed what could be generated artificially by induced mutation, and it is this pool that plant geneticists most often exploit. Modern agricultural practices, however, concentrate genetically uniform, high-yielding varieties in massive pure stands from which weeds, including weed relatives, are ruthlessly excluded. Maize provides a striking example. In the wild, it hybridizes with a weedy relative, teosinte. This is one source of new genes; however, teosinte is rapidly being eliminated as a companion weed. Wilkes estimates that the distribution in Mexico and Guatemala of teosinte is today only about one-half the area it occupied one hundred years ago. The potato is another case in point. Many of the wild varieties of potato growing in America when the Spaniards arrived have become extinct. Their genes are probably lost forever. Man has settled on a few high-yielding varieties to the exclusion of others.

In such species as modern corn little variability is left in the crop plant because the development of a hybrid by manipulated cross fertilization establishes a genetically uniform stock. Corn is one of the plants that can be either self-pollinated or cross-pollinated. If corn plants are self-pollinated for a number of generations, this inbreeding eventually produces weak, runty plants. The evils of inbreeding were well known by European royalty a century or more ago and by certain isolated American rural communities where human inbreeding yielded similarly handicapped offspring. If, however, two inbred strains are crossed, the hybrid exhibits unusual vigor, which may even exceed that of the original parental stock from which inbreeding started. In the case of corn, a double cross involving four inbred varieties is made. In the first year the unopened tassels (male flowers) of every plant of one variety are removed so that the plant cannot pollinate itself. Its ears (the female flowers) are pollinated by pollen from a different inbred strain that is growing nearby and has not been detasseled. Few seeds are produced. These are sown the following year and produce a vigorous hybrid with many seeds. At the same time, another hybrid has been produced by crossing two other inbred strains. Now with the two hybrids the crossing is repeated once again; one is detasseled and pollinated by the other. Many seeds are produced and these are the ones that are employed in the third year for the final harvest. The production of hybrid seed is a very specialized industry that can make seed to meet almost any specification of climate, soil, and susceptibility to disease.

The dangers inherent in setting up monocultures of genetically uniform stock have recently been pointed out in a report of the

National Academy of Sciences, which has established a committee to determine how narrow is the genetic base in crops. When a hardy variety of plant, resistant to the multiple ills affecting crops, has been developed by cross-breeding and selection, as in the case of corn, farmers naturally plant this hybrid in preference to others. Such popularity can result in the nation becoming dependent on a single, highly-selected variety of plant. The risk inherent in this practice is the potential vulnerability of the genetically uniform stock to new types of pathogens.

In nineteenth-century Ireland, the cultivation of potatoes was restricted almost entirely to the Lumper and other high-yielding varieties. These happened to be particularly susceptible to fungal attack. Other varieties that were resistant were restricted to small local plots. One of these was the American Early variety; but before this variety could be introduced into potato farming as a major crop, a million Irish had died from starvation and another million had fled the country. Modern agriculture may be more enlightened in this respect, but the risk is still there. That the risk is ever present was devastatingly demonstrated in the summer of 1970 when the Southern Corn Blight struck the American crop without warning. The subsequent loss was estimated at a billion dollars.

The fungus that causes Corn Leaf Blight in the past has been responsible for a two or three percent loss of the corn crop in the United States. The popular corn hybrids were resistant to it. Then within a single season strains of hybrid corn genetically engineered to have a type of male sterility called T cytoplasmic male sterility were devastated by a new strain of the fungus. The purpose of developing male sterility was to save the labor costs of detasseling plants by hand. Susceptibility was clearly associated with the particular type of cytoplasm. From Florida, the spores moved through the Gulf States and then with prevailing winds up to the middle Atlantic and midwestern states and eventually into Canada.

When man introduces into nature new varieties of plants that are resistant to a particular pathogen, he upsets the conditions under which that population of the pathogen exists. The population is not homogeneous; it consists of many strains of the pathogen. The ratio of the strains depends upon the kinds of host available. As long as the host population is made up of a particular susceptible variety, the strains of pathogen that can live upon it are in the majority. Those strains that normally lived on a susceptible variety of host are reduced in number as the new resistant variety of host plant becomes the one most popular with farmers. Then strains that normally constituted a minority increase. Furthermore, since pathogens are constantly mutating or otherwise genetically changing, the new ones increase when new susceptible hosts become available, while others decline in numbers. Virulent forms can be present in small numbers for years, merely awaiting an opportunity to explode. This happened with a wheat stem rust, the so-called race 15B, which for at least 11 years occurred in low numbers in the United States. By 1953, it had changed enough genetically to become the major rather than minor component of the rust population. In 1953, it destroyed 65 percent of the durum wheat crop; in 1954, 75 percent of the durum wheat crop and 25 percent of the bread wheat crop. The geneticist Tatum has pointed out that evolutionary processes in the heterogeneous populations of pathogens mean that a new variety of host crop resistant to a particular strain of pathogen will eventually become susceptible.

The second consequence of domestication is the drastically altered environmental setting in which crop plants grow. Part of the ability of plants to survive the onslaughts of herbivores and parasites lies in their health and the physical conditions to which they are subjected. The former also depends in part on the latter, so the physical setting acts in a dual capacity. Even an amateur gardener can appreciate what this means.

In cultivating plants, man prepares the soil for them, alters their diet by manuring, regulates moisture settings by mulching, provides water on demand by irrigation, removes competition from other plants by weeding, offers protection from such obvious harm as trampling by large animals and attack by birds, shapes their growth by pruning, sets aside and protects their seeds in times of adversity, sows the seed or sets out the cuttings at propitious times, thus avoiding climatic challenge. In greenhouses, he even regulates length of day. In some orchards (e.g., citrus groves) he actually neutralizes climatic extremes by putting out smoke pots. He sprays on various growth substances to modify growth, to reduce or increase amount of foliage, to prevent fruit drop. He creates monstrosities by grafting. All of this he does by design. Unwittingly he effects other changes, too, as, for example, polluting the air so that the plants suffer from excessive ozone, or growing plants under conditions where subtle deficiencies of growth micronutrients cause defects. The invidious connotation of "hothouse" when applied to human activity clearly reflects some of the less desirable consequences of pampering.

Most obviously man has altered plant populations in much the same way that he has altered his own. He has concentrated dense populations of single species. In America, in 1930, fifty million acres of wheat were under cultivation. A wheat field may stretch from horizon to horizon. In the Sudan, nearly a quarter of a million acres of cotton may cover one continuous area. One of the consequences of large uniform plantings is a genetic one. We have already discussed briefly some effects of direct interference with genetics. Huge monocultures provide examples of indirect interference. In the crops themselves, in some cases, variability may persist; but it will be of a different kind than that found in wild types.

One of the normal evolutionary processes that has been radically altered by man's development of monocultures on a grand scale has been that of selection. Normally among wild plants, small isolated populations favor a combination of chance divergence and adaptive changes leading to the development of races and eventual speciation. Cultivation of such plants as wheat, cotton, and peanuts, to mention only a few, involves fields comprising thousands of acres of pure stands separated only by fences, ditches, and highways. One of the consequences is loss of isolation. There is no privacy, so to speak, so that there is a free exchange of seed and interbreeding on a vast scale. The result is genetic uniformity. Add to this the ease of intercourse that modern transportation has made possible, and it becomes clear that the chance of isolated populations establishing themselves becomes exceedingly remote. The selective forces that operate on these monocultures are very different from those working on wild species. Weeding and cultivation eliminate interspecific competition, which is then replaced by intense intraspecific competition. Among the changes in variability that this induces is a loss of those characters that are needed for surviving in natural mixed vegetation. In the case of cotton, certain wild characteristics (drought resistance, intolerance of shade, and competition among seedlings) have persisted in highly-advanced cultivated varieties whereas others (hard-seed coat, delayed germination, the perennial habit, facultative shedding, and photo-periodic responses) are lost.

Some people do not regard specialization in crop plants as a prelude to evolutionary disaster. They argue that the wide range of ecological conditions which local strains have been developed to meet, the great size of interbreeding populations, and the extent to which modern transportation has permitted exchange between widely separated areas minimize the chances of irrevocable loss of valuable genetic material. In support of this contention they point to the success in discovering and re-establishing varieties of genes for disease resistance in certain crops—genes which, long neglected, have persisted in domestic populations. For most plants, this ghetto or botanically urban way of life is unnatural. In nature, these conditions are approximated only by extensive stands of grasses in prairie and savannah (but even here the stands are not pure) and in the taiga, the vast coniferous forests that extend in a wide circumpolar belt throughout the northern reaches of the Old and New Worlds.

Thus, through domestication man has not only produced freaks but he has grown them under unusual conditions. It might be argued that while these conditions are inimical to the well-being of wild plants they are Utopian for domesticated ones. This may well be true. Many cultivated plants look healthier, more robust, more productive than their wild counterparts. Were this the only consideration the whole argument would be pointless. But these plants, beautiful to behold, loaded with a surfeit of excessively large reproductive organs, may be ideally developed to live in their artificial environment, were this environment an isolated cosmos, a closed system. Unfortunately, these plantations are no more isolated from the world at large than are our cities from the rest of the country. There are powerful extraneous forces constantly at work.

The impact of these forces is nowhere more apparent than when either the plants or the forces are tranplanted to alien shores. It is in this context that the impact of dispersal is seen. It is axiomatic in agriculture that crops are grown most successfully away from their areas of origin. The reasoning goes that by removing plants to foreign climes one enables them to escape from their pests and parasites. (This assumes that their pests are not stowing away on them.) One might argue that a plant would be in better evolutionary balance with the pests in its own neighborhood than when exposed to alien infections. Were the domesticated plants the products of such evolutionary interactions rather than of the irrelevant manipulations of man this might well be true. There is, however, another way of looking at the situation which becomes clear when one considers dispersal of the other members of the partnership.

When a herbivorous insect is dispersed from its point of origin, it frequently becomes a pest of formidable significance. In considering the dispersal of plants and their pests, movement is not relative. The absolute directions count. A plant moved to a new geographical area may well be strong and free of the insect pests that evolved with its ancestors at the point of origin and are only too willing to transfer their attentions to the effete and susceptible offspring. But moving the insect to a new geographical area does not mean that plants there are immune to it because these are not normally associated with it. What it usually means is that the insect is freed from its own pests and parasites and can expand its population in health and safety just as the transplanted plant did. Thus, there are fewer curbs on its appetites, and it can undergo a population explosion to the detriment of the weak domesticated plants and native plants that have not evolved defenses.

Thus, the gypsy moth was able to reproduce with impunity in the New World and defoliate acres of forest in New England. Similarly, the Hessian fly that came to America during revolutionary times had a field day in wheat. And if the agriculturalist is so unfortunate as to have the crop that he carefully imported to a safe part of the world followed by one of its former insect associates, a catastrophe can result. Potatoes grew well in Europe until the Colorado Potato Beetle, its home-town nemesis, followed it there. The devastating effects that some imported insects have on plants suggests that back in their own bailiwick they were being limited not by the finiteness of the food supply but rather by the check imposed by climate or by predators and parasites.

In domesticating plants, man has also unwittingly cultivated insects. He has made conditions for their growth and propagation more favorable. By growing pure stands of plants, and plants that are susceptible to insect attack by having been removed, as it were, from the free competition of evolution, he has provided unlimited food and brand new environments. By extending the growing season and alternating crops, he has provided continuous supplies of food where previously the supply was intermittent. In Israel, two generations ago, one summer cereal, Durra (*Sorghum vulgare*), was sown in May and was dry by July. One of its pests, the sorghum borer (*Sesamia cretica* Lederer), could produce one generation, occasionally two, on this crop. Today, because of irrigation, several varieties of sorghum and maize are green and luscious from April to November. The sorghum borer can now raise three generations!

Man's establishment of huge monocultures in areas that were formerly rich in diversity has resulted in a limited number of insect species concentrating in great numbers on one plant instead of small numbers of many species distributing themselves among many plants. In one example from the steppes of Kasakhstan, the virgin land supported 312 species of arthropods whereas an adjacent monoculture of wheat supported only 142 species. In the wheat field, however, the population density was three times that of the steppe, some species characteristic of the steppe had disappeared, specific wheat insects had emigrated to the wheat from distant fields, and insects that were common to wheat and to steppe increased more rapidly in numbers on wheat and became from 280 to 550 times more numerous.

Man engineered plant population explosions and opened the way for insects to increase too. When he abandoned a crop or a particular kind of crop fell out of favor, the insect population must also have declined. Man further exacerbated the insect-plant relationship by attempting to unbalance the prey-predator aspects of the situation. Under normal cicumstances, the insects and plants left to themselves would eventually have achieved a new balance. Thus, in the great coniferous forests of the taiga the introduced spruce budworm, released from the pressures of its native parasites and predators, would have killed millions of trees; but then the reduced acreage of trees would eventually impose a restraint on the insect population explosion. A fluctuating balance would result.

Periodic outbreaks in natural ecological situations are a normal feature of nature. They help insure diversity and stability. They tend to occur in unstable situations, for example, in arid plant communities, and in undiversified systems, as, for example, in simple coniferous forests. These are the features of cultivated systems—instability and lack of diversity. Outbreaks have never been recorded from a mature tropical rain forest, which is both diverse and stable.

In Maine, in 1818, every spruce tree west of the Penobscot River was killed by some unidentified insect or insects. In 1881, it was reported that the spruce forests had completely regenerated. Similar catastrophic destruction of spruces was recorded between the years 1878 and 1881 in northern New England. Again in its own good time the forest returned. There has also been cyclic destruction and regeneration of forests in Germany and Scandinavia.

Outbreaks in forests never arise simultaneously over large areas. A population will begin to increase in some favorable center, spread out over the course of several years, and then die out. Favorable centers are usually provided by deterioration in the tree population—old trees, excessively dense populations, wind or fire damage—and favorable weather conditions. It has been shown that definite patterns of atmospheric circulation characterize the years preceding outbreaks of spruce budworms and forest tent caterpillars. Ultimate collapse of the outbreak is caused by competition for food and space, a build-up of parasites and predators, unfavorable weather, and a genetic deterioration of the population.

Man, however, tends to think in a time scale geared to his own life span. The immediate consequences are more important to him than the long-term ones. And since he wants his timber now in a specified quantity, he is not willing to suffer the temporary loss. Therefore, he artificially maintains a high plant population, and the insect has no chance of eating itself out of food, house, and home.

Although a superabundance of food may allow an insect species to increase in numbers, and a decrease in food cause a decrease in numbers, food does not tend to be the major limiting factor. Rather it is a contributory one. Considering the botanical side of the relationship, although insects may wipe out a plant population locally, as for example, 100 percent consumption of a corn crop by locusts, this tends to be the exception rather than the rule. Gypsy moth caterpillars may completely defoliate a forest, but the trees grow a second crop of leaves. The trees are admittedly weakened by this attack and rendered more susceptible to other infestations, but the forest will grow back. We shall return to this in Chapter Two.

MAN-MADE ECOLOGY

To complete this rough sketch of the effects that man has had on plant-insect relationships, we must look briefly at the indirect consequences arising from the simple fact of his making areas habitable for himself. Without quite knowing what he was doing through the centuries, he has redesigned vast ecological panoramas. The work continues all about us, but the sweeping changes on the face of the earth can only be appreciated from an historical perspective because little of the original communities of plants and animals remains to afford a comparison.

The most obvious alteration has been deforestation. Everyone can appreciate the difference between a forest and the absence of a forest; however not everyone is aware that deforestation is more than the mere removal of trees. It involves the removal of all the animals that once lived there, the introduction of new species of plants as well as animals, changes in the conditions of the soil, water relations, and climate.

Consider what happened in North America. In the northern states, during the 1600s and 1700s, human activity outside of the colonial villages consisted principally of hunting and trapping. Demands made on the forests were few. It was not until the 1800s that harvesting of lumber reached full stride. When the colonists first ventured into the forests, they found a patchwork quilt of trees, stands of hardwoods with a few conifers, and stands of conifers with a few hardwoods. In the beginning, in the 1700s, hardwoods were cut selectively: hemlock for its bark for tanning, maple and cherry for furniture, yellow birch for veneer. Clearing by fire was infrequent, and vigorous species like maple and beech regenerated quickly by resprouting. In the patches of conifers the story was different. At first slowly, then at an accelerating rate, stands of white pine were obliterated. Lumbermen started in Maine in the 1700s. As the white pines were mowed like so much grass, the axe advanced on New York (1850), Michigan (1870), Wisconsin (1880), and Minnesota (1890). As much as 90 percent of the white pine was removed. Left behind was desolation, slash, and burns.

White pine is a "fire" tree; that is, it is adapted for seeding itself after a burn. Since it cannot sprout from a burned stump, the only source for a new population is seed; however, if a second fire or other catastrophe intervenes before the first survivors have reached maturity, there is no seed, and no more pine forest. Aspens, birches, oaks, and hazelnuts replace the vanished conifers.

In a more general sense, opening the virgin forests causes changes of great complexity. The microclimate changes; there is an increase in evaporation, the habitat becomes drier. The plants whose seeds are carried into and succeed in this newly made space are just those adapted to the drier conditions. The original mixed forest is replaced by great expanses of aspens, scrub oak, box elder, and sassafras, which with repeated harvesting become the new permanent inhabitants.

Deforestation is only one example of the magnitude and character of change that can take place. Not even the grasslands of the mid-continental Great Plains were immune. Before the coming of the Europeans, they were occupied only by Indian hunters. The main herbivores were bison, pronghorns, rabbits, prairie dogs, ground squirrels, gophers, and kangaroo rats. In the sixteenth century, the plains probably consisted of tall prairie grass to the east and short "plains" grass mixed with weeds (forbes) and woody species in the west, grading into desert shrubs farther to the west, southwest, and northwest. Little remains of this native flora except along railroad rights-of-way which were often protected by fences and burned no more often than in prehistoric times. The plains were invaded by so many foreign grasses and forbes (dandelion, white clover, ox-eye daisy, quack grass, and Timothy) that they resemble a European man-made community. The original plains were in flucuating balance with their native herbivores. It has been estimated that in one period, California ground squirrels ate plants which could have supported 160,000 head of cattle or 600,000 sheep. Drought was probably the severest pressure to which the grasslands were subjected. The periodic invasions of locusts became critical chiefly when combined with drought and grazing by rodents.

The man-made plains and prairies were then subjected to new pressures. Millions of horses, cattle, sheep, and goats were turned loose to share the grazing with the native rodents and locusts. The horse was the first invader, having spread in the seventeenth century by escape and Indian capture from the Spaniards in Santa Fe. By 1890, the horse, by providing mounts for such "heroes" as Buffalo Bill and others of his kind, had nearly eliminated the estimated fifty million bison that together with the pronghorn were the original ungulate grazers. The horse was joined by hundreds of thousands of cattle spreading from Texas up through

Kansas. How influential all these invaders were in changing the character of the prairies is debatable. Certainly, there was local overgrazing and cropping of grass throughout its reproductive stages. As a result, new species of plants invaded the area. As with the forest, there was a shift from the more permanent (climax) species to pioneering (transient) species. There was also a shift to generally more arid conditions.

In addition to all these influences came fences, hedges, wells, irrigation ditches, houses, and towns. The final wound on the prairie was delivered by a steel plow strong enough to bust the tough prairie sod—a plow invented and manufactured in the mid-nineteenth century. Cropland obliterated grassland. The corn belt of today is the man-made substitute for a grassland that extended in post-glacial times into bordering forests and was maintained in that state by recurring fires.

The changes that have taken place since the sixteenth century have visibly altered the face of the middle of North America and produced hidden but drastic effects. Soil stability decreased, novel sub-soil water conditions arose, the reflective power of the ground (albedo) and hence the temperature changed, and a trend toward drier, lighter, and more variable conditions was established.

In North American forests and prairies, within a space of time that is only a tick of the clock of evolution, man created, as he has done throughout most of the world, a whole new ecology. It is in this new world, probably as new as any produced by slowly acting geologic forces, that plants and insects have suddenly been thrust and must pursue their co-evolution.

To a greater or lesser extent, depending on individuals and circumstances, we all tend to ignore change, to fight change, or to wish it away. As Santayana soliloquized: "Change to us is an omen of death, and only in the timeless can we feel secure." When we face up to reality, however, we realize that change is a part of the essence of the universe. It is hardly surprising, therefore, to discover that populations of insects change also, both in quantity and quality.

In stating earlier that insects and plants declared a truce over one hundred and thirty million years ago no impression was made that the numbers of each species were constant. Instead, we described a situation in which plants and insects were in a state of equilibrium such that neither prey (the plant) nor predator (the insect) became extinct or excessively abundant. We implied that the population of each insect species was stabilized at any given period of time so long as the conditions affecting population change remained relatively constant. In the long evolutionary view, the essence of the insect-plant relationship is one of episodic dislocations which make for an uneasy equilibrium. When man finally appeared on the scene he introduced changes that upset stabilized relationships on a worldwide scale. Today, we can see populations that are obviously stabilized and others that are in various stages of readjustment to new conditions, re-stabilizing to a higher population level in some cases, and to a lower level in others. Some are moving inexorably towards extinction—mainly, but not exclusively, as a result of man's modification of environmental conditions and their inability to adapt to change.

The situation is more obvious to the layman if he observes large animals. For example, in the eastern United States the population of white-tailed deer has increased since Indian times because European colonization, by its agricultural practices and urbanization, created more cleared land. In response to these changes, timber wolf numbers decreased nearly to zero. In the case of birds, the introduction of the English Sparrow resulted in a population explosion that was more or less stabilized until the replacement of the horse by the model T Ford caused a readjustment downwards to the present population density. Similar movements occur among insect populations. How the changes that man introduced affected insect populations is best appreciated by considering very briefly the factors that bring about changes in populations in general.

INSECT POPULATIONS

Every species of animal has an intrinsic capacity to increase. Thus, a female tsetse fly throughout her lifetime of six months can produce twelve young, two per month. A housefly, on the other hand, can produce during her average lifetime of two to three weeks seven hundred and twenty eggs. The rate of increase is implemented not only by how long is required to produce each egg or young but also by how long it takes the female to reach reproductive age. A tsetse fly, for example, requires about fifty days from the time she begins independent life as a freshly laid larva herself till she produces her own first larva. A housefly requires from ten to fourteen days from hatching to oviposition. An American cockroach may require as long as one year. The seventeen-year locust requires seventeen years. The absolute numbers here are determined by the individual physiology of the insect concerned, by the intrinsic characteristics of growth, development, and reproduction. These are genetically determined.

Actual reproductive capacity as compared to potential reproductive capacity is modified directly and indirectly by many factors. Weather is an important modifier. Excessive cold can delay or impair reproduction. Shortage of food or food of inferior quality can reduce birth rate or the viability of eggs and young. Anything that impairs the health or longevity of the female or the opportunities to find a mate or sites for deposition of eggs (as, for example, crowding) can reduce her output. Last but not least, anything that causes her death obviously affects her capacity to reproduce. In practice the measurable result of these many interactions is the birth rate and the death rate.

Insofar as the population of any given species is concerned, the rate of increase in numbers in an Elysian environment would be determined solely by the intrinsic characteristics of the species. It is the genetic background that determines such demographic occurrences as mating, birth, death, and dispersal. It is the contemplation of intrinsic unchecked reproduction that enables one to calculate the staggering inexorable march of protoplasm as, for example, Darwin did for elephants. Assuming that a pair of elephants started breeding at 30 years of age and lived to be one hundred years old, there would be after 750 years approximately nineteen million elephants!

In the mortal world, however, this infinite exponential growth never occurs because negative factors take a directive hand. These are the environmental factors that prevent birth rate from achieving its maximal potential and also accelerate death rate. Obviously some demographic occurrences are modified more by environmental influences than others. Birth rate is primarily a matter of species characteristics and only modified by environment, while death rate is primarily a matter of environment and only modified by species characteristics. The most decisive environmental events include climatic ones, disease, predators, parasites, accidental death, desiccation, and starvation. These may be independent of the size of the population or may be brought about as a direct consequence of a dense population.

Density of population influences the number of predators and parasites, the incidence of disease, the vitality of individuals, their longevity, the amount of food available, and any number of behavioral characteristics that affect mating and reproduction. In any given time and space, the actual numbers of a species depend not only on the interaction between birth rate and mortality but also on the movement of the population. Thus, anything that affects migration and dispersal or the failure to move affects the local density of a population.

In a man-less world, the fluctuations of populations of insects over long periods of time are characteristically different for different species. This is because the characteristics and relative contributions of the co-determinants (primary or demographic and secondary or environmental) affecting population size differ from one species to the next. Certain grasshoppers and locusts, as well known by the ancients, fluctuate wildly from one extreme to another. Other insects, of which the black-headed budworm in New Brunswick, Canada, is an example, fluctuate in a predictable manner about a mean. The grasshopper population is determined principally by weather, a phenomenon that operates independently of the density of the population. The budworm is dominated by its parasites, which are proportional in number to the number of budworms. Generally speaking, the insects that are prone to outbreaks are those in unstable conditions. And weather is probably the commonest disturbing factor.

Man's interest in the plant-insect relationship is by no means impartial. He wishes to redress balances that are intrinsic to the system irrespective of whether the system is left to itself or, as is more often actually the case, is deranged by his action. His basic problem is that he wishes to perpetuate a stable imbalance in favor of certain plants (his crops) in a biological system that is not static but in fluctuation. He not only desires static imbalance in what is fundamentally a fluid situation, but he wishes immediate results. In short, agriculture is an artificial phenomenon marching to man's cultural tune in an evolutionary parade that marches to a biological tune.

References

1. Agrios, G. N., *Plant Pathology*. Academic Press, London and New York, 1969, 629 pp.
2. Anderson, E., *Plants, Man and Life*. University of California Press, 1967, 251 pp.

3. Baker, H. G., *Plants and Civilization*. 2nd ed. Wadsworth Pub. Co., Belmont, California, 1970, 194 pp.

4. Baranyay, J. G., Lodgepole pine dwarf mistletoe in Alberta. Department of Fisheries and Forestry. Canad. Forest Ser. Pub. 1286 (1970), 1-22.

5. Brooks, C. E. P., *Climate Through the Ages*. Revised edition. Ernest Benn Limited, London, 1949, 395 pp.

6. Clark, G., *World Prehistory. An Outline*. Cambridge University Press, 1965, 284 pp.

7. Clark, L. R., Geier, P. W., Hughes, R. D., and Morris, R. F., *The Ecology of Insect Populations in Theory and Practice*. Methuen, London, 1967, 232 pp.

8. Coon, C. S., *The Story of Man*. Alfred A. Knopf, New York, 1954, 438 pp.

9. De Candolle, G., *Origin of Cultivated Plants*. D. Appleton & Co., New York, 1885, 488 pp.

10. Deevey, E. S., The human population. Sci. Amer., Sept. 1960, pp. 198-205.

11. Essig, E. O., *A History of Entomology*. MacMillan, New York, 1931, 1029 pp.

12. Evans, E., *Plant Diseases and Their Chemical Control*. Blackwell Science Pub., Oxford, 1968, 288 pp.

13. Good, R., *The Geography of the Flowering Plants*. 2nd ed. Longmans, Green and Co., London, 1953, 425 pp.

14. Graham, S. A., Ecology of forest insects. Ann. Rev. Ent., 1 (1956), 261-80.

15. Harlan, J. R., Agricultural origins: Centers and noncenters. Science, 174 (1971, 468-74.

16. Iverson, J., Forest clearance in the stone age. Sci. Amer., March 1956, pp. 22-27.

17. Large, E. C., *The Advance of the Fungi*. Jonathan Cape, London, 1940, 488 pp.

18. Leopold, A. C., and Ardrey, R., Toxic substances in plants and the food habits of early man. Science, 176 (1972), 512-14.

19. MacNeish, R. S., The Origins of New World civilization. Sci. Amer., Nov. 1964, pp. 13-21.

20. Morris, R. F., Observed and simulated changes in genetic quality in natural populations of *Hyphantria cunea*. Canad. Ent., 103 (1971), 893-906.

21. National Academy of Sciences, National Research Council. The Plant Sciences Now and in the Coming Decade. Washington, D.C., 167 pp.

22. National Academy of Sciences. Washington, D.C. News Report. March 1971, 21 (3), 2-3.
23. Osborn, H., *A Brief History of Entomology*. Spahr and Glenn Co., Columbus, Ohio, 1952, 303 pp.
24. Rivnay, E., The influence of man on insect ecology in arid zones. Ann. Rev. Ent., 9 (1964), 41-62.
25. Sauer, C. O., *Agricultural Origins and Dispersals*. Amer. Geographical Soc., New York, 1952, 104 pp.
26. Santayana, G., *Soliloquies in England and Later Soliloquies*. University of Michigan Press, Ann Arbor, 1967, 264 pp.
27. Simmonds, N. W., *The Evolution of the Bananas*. Wiley, New York, 1962, 170 pp.
28. Tatum, L. G., The southern corn leaf blight epidemic. Science, 171 (1971), 1113-16.
29. Ukers, W. H., *All About Coffee*. The Tea and Coffee Trade Journal Co., New York, 1935, 818 pp.
30. Usinger, R. L., The role of Linnaeus in the advancement of entomology. Ann. Rev. Ent., 9 (1964), 1-16.
31. Vavilov, N. I., *The Origin, Variation, Immunity and Breeding of Cultivated Plants* (transl. K. S. Chester), Ronald Press, New York, '1951, 364 pp.
32. Von Loesecke, H. W., *Bananas*. Interscience Pub., New York, 1949, 189 pp.
33. Weed, C. M., *Insects and Insecticides*. C. M. Weed, Hanover, N. H., 1891.
34. Wilkes, H. G., Too little gene exchange. Science, 171 (1971), 955.
35. Wilkes, H. G., Maize and its wild relatives. Science, 177 (1972), 1071-77.
36. Wilson, C. M., *Empire in Green and Gold*. Henry Holt, New York, 1947, 303 pp.
37. Wilson, C. M., *Grass and People*. University of Florida Press, Tallahassee, 1961, 233 pp.
38. Wilson, F., The biological control of weeds. Ann. Rev. Ent., 9 (1964), 225-44.
39. Ziegler, P., *The Black Death*. Penguin Books, Harmondsworth, 1969, 331 pp.

Chapter Two
The Third Horseman

2
The Third Horseman

Two "revealed" truths appear to permeate all contemporary thinking about the status of insects in the ecological triumvirate man-plant-insect: The first is that the insect actually and potentially is the most formidable and implacable competitor of man for a finite amount of vegetation; and the second, that the role of the insect is becoming increasingly dominant. Both of these allegations warrant careful scrutiny. In this chapter we shall analyze them in an historical context. In the following chapter we shall attempt a contemporary assessment.

Famine has been a recurring theme in the Old World. In 436 B.C., famine was so severe in Rome that people drowned themselves in the Tiber. A great famine visited Egypt in A.D. 42. In 650, 879, 941, 1022, and 1033 famines of historic proportions depopulated entire provinces in India. Other great famines occurred in England in 1005; throughout Europe in 1016; for seven years in Egypt (1064-72); for eleven years in India (1148-59); throughout the civilized world in 1162; again in India in 1344-45; for twelve years in India in 1396-1407; in England in 1586; in India in 1661, when not a drop of rain fell in two years; in India periodically to the present; in Ireland in 1846-47; in China in 1877-78, and again in 1887-89; in Russia in 1891-92 and in 1905. The events leading to famine differ with time and place. Let us choose for examination an area where extensive documentation is available. Let us look back in time to events that occurred in the Old World (in Europe, and particularly England) as they affected the sustenance of man.

With the exception of parts of the Near and Far East, Europe has been under continuous intensive cultivation longer than any other part of the world. Before the advent of man, the whole subcontinent was clothed in forest, dense massive oak forests in western and central areas mixed with beech and conifers in mountainous sections, a vast coniferous belt in the north, both west and east, and an open evergreen forest of oaks and pines in the south along the shores of the Mediterranean Sea. Clearing began almost as soon as man arrived on the scene. The rate was accelerated when the polished stone axe replaced the less efficient chipped axe. Throughout successive centuries, the forests waxed and waned like a great green tide as areas were abandoned and reclaimed. When in the eleventh century the Germans advanced southwest and northeast as colonists, much as the American pioneers advanced westward from the Atlantic, they found the furrows of long lost and abandoned Saxon farms. In southern Europe, there are even historical accounts of the destructions of the forests. By Classical times, much had already disappeared, and because of the poor regenerative powers of the original evergreen species, such forests had been replaced by scrub and bare soil. In the ninth century B.C., Homer had alluded to forests ("wooded Samothrace"), and Plato recorded the destruction of forests in Attica to provide timber for the construction of ships and houses.

But Europe as a whole is too vast and varied to serve our purpose best. It has markedly different climatological regions and different and unequal rates of agricultural development. Furthermore, sources of information relating to the continent as a whole are frequently incomplete for critical periods. It better serves our purpose to focus on one particular area where extensive documentation is available. England fulfills all our requirements, but even in this case we must keep in mind the fact that bias in documentation requires careful interpretation of statements. Some compilations were concerned with law, some with taxes and tithing, some with ecclesiastical holdings, some with manorial matters, etc. The original purpose of a document, whether it dealt with legal, economic, sociological, or ecclesiastical matters, determined what was mentioned and what was overlooked. An omission meant very little.

AGRICULTURE IN ENGLAND

Agriculture arrived late in England, but documentation there was probably more complete than anywhere else in the Old World; consequently, the agrarian economic history is well-known.

Evidence relating to conditions between the years A.D. 450 and 1086 can be gleaned from place names and from the Domesday

Book which recorded forest assets. In the southern county of Middlesex, on light gravel and loam soils, there are many names ending in *-ham* and *-cote*. They do not indicate the presence of woods, and belong to early phases of Saxon settlement when land was already cleared. Around London, on the other hand, the soil was never so arable, and place names indicating woods persisted for a longer time. There are no ancient names from the northeast because it was at that time still a vast unsettled woodland. Records during the early Middle Ages were maintained principally by the monasteries, which represented the great farming interests of that era. Later came the records of the great private landlords, whose principal interests were the collection of rents and the control of leases. Accordingly, the point of view from which man's relation to plants is viewed is a constantly changing one; nevertheless, a broad picture can be faintly discerned through the mists of antiquity, clear enough to allow us to see what forces were at work in man's tug-of-war with soil and climate.

When Paleolithic and Neolithic hunters and gatherers roamed the British Isles between 300,000 and 3500 B.C., the islands, like the rest of Europe, lay under green blankets of dense forests. There were no grassy, rolling downs, no barren Scottish highlands carpeted in heather, no "natural" heath. Instead, there were dark forests inhabited by wild boars, spirits, ogres, and lesser deities. These forests were a source of fuel and charcoal and eventually pasturage for swine (which ate acorns). Around 3500 B.C. immigrants from the Continent introduced agriculture. Agriculture had already attained high levels of development in the Near and Middle East, but its advance into Europe was slow.

The earliest authenticated agricultural culture was the Neolithic Danubian flourishing north of the Danube between 5000 and 6000 B.C. This was the ancestral home of Britain's first farmers. Their principal crops were barley, einkorn, emmer, possibly supplemented by bread wheat, beans, peas, lentils, and flax. The culture as it came to Britain was a hoe culture that relied on slash and burn techniques to clear the forest. Clearings were immediately invaded by scrubby pasture, which was burned and sowed with cereals. Settlements were of short duration because the fertility of the soil was soon exhausted. Re-colonization of abandoned fields occurred only after the forest had regenerated.

Some idea of what plants grew at what times and how they

succeeded one another has been gained by studying pollen preserved for thousands of years in the soil. Under the microscope,
pollen grains of each species of plant are seen to be distinctively
and characteristically sculptured. Each can be identified by comparing it with contemporary grains. In England, from evidence of
pollen analysis and plow marks left as scowering on soft bedrock,
we surmise that wheat and a primitive form of barley were grown.
Cows, sheep, and goats were kept and wild pigs were eaten; however, the mainstay of life was game and wild plants. The men of
that time were the people who built, probably as early as 2500 B.C.,
the first of the three temples that make up Stonehenge.

A second European culture reached England between 2000-
600 B.C. These people, a Bronze Age culture known as "the
Beaker people" after the shape of their pottery, built the second
temple at Stonehenge between 1600-1500 B.C., and the third and
final temple at the end of the Early Bronze Age, about 1400 B.C.
By the beginning of the Middle Bronze Age (1400-1000 B.C.),
they, or those who followed them from the mainland, had introduced the practice of manuring. Two new kinds of winter grain,
einkorn and husked barley, were also introduced sometime between 1000 and 500 B.C. From the end of the late Bronze Age
and the beginning of the Iron Age (600 B.C.), until the Belgae
arrived in 100-50 B.C. and Caesar in 55-54 B.C., these Celtic
people, the Britons as they are now called, grew on their small
square farms enough corn (i.e., wheat), barley, flax, and woad
(dye) that they could actually export grain.

During the halcyon days of Roman occupation, from A.D. 43 to
A.D. 409, when the Romans left the Britons to the mercies of the
Picts, Scots, Angles, Jutes, and Saxons, there was little change in
agricultural practice. It was a cereal-based agriculture more than
ample to the needs of the population. The Saxons drove the hapless
Britons into the vastness of Wales and Cornwall, where King
Arthur waged a losing battle for survival, the Vikings invaded and
intermingled in the eighth century, and the Normans took over in
1066. Through all these changes and troubled times the agricultural situation remained essentially unchanged and sufficient
for the needs of the inhabitants.

With the ushering in of the Middle Ages difficulties arose. What
Saxon England began—namely, the concentration of populations
into towns and a gradual change of agriculture from subsistence

farming to commercial farming—the Normans intensified. Urbanization had a profound effect on supply and demand. As trades and specialities began to develop, even though every man did some gardening, the commercial farmer evolved. Consideration of these developments is most important in evaluating man's relation to plants. Until that time, each man raised what he needed and was even able to accumulate something from the bountiful years to tide him over the lean years. Supply kept up with demand. But medieval economy was no longer a subsistence economy, even though many people lived at a subsistence level. At the same time, the population began to grow. It rose rapidly from the tenth century until about 1300. Bad harvests, from 1317-19, due mostly to bad climate, brought famine. Between the years 1348 and 1350, the Black Death together with war caused a catastrophic decline in population. It was not until the second half of the fifteenth century that it began to climb once more.

What was the agricultural situation of the times? In the tenth century grain was *the* crop. Wastes and forests, the former homes of the early slash and burn agriculturalists and their flocks, were replaced by permanent ploughed fields. In the tenth and eleventh centuries, the control of land and of the peasants who worked it was exercised by the seigneurie, that is, the heads of monasteries, the princes of the Church, secular princes, war lords, military aristocracy, and minor men of affairs. The holdings of these landlords were divided into the demesne, that land farmed exclusively for the landlord, and the tenements, those parcels of land farmed for the partial benefit of the peasants. At first, grain was the principal crop, but it gradually yielded in some areas to profitable commercial crops, to supply demands created by a rise in aristocratic and urban living standards in the twelfth and thirteenth centuries. Cash or its equivalent was to be had for grapes (wine), flax (linen), hemp, dyestuffs, and wood. Grubbing of forested lands intensified during this period of great rural conquest.

The rise of a consumer market encouraged gardening at the expense of farming. Now there was a more rational exploitation of forests and more fencing. In the thirteenth century, bacon, herring, and cheese became more important items of diet. But the expansion had reached its highwater mark. The urban economy had attained its maximum prosperity; depression ensued. Between 1300 and 1440, many complex factors interacted to bring on a crisis that included an ebb in agriculture. War, pestilence, a decline of seigneurial estates, and opppressive taxation dealt severe blows to agriculture. The taxation was so onerous that even the down-trodden peasant could tolerate no more. In 1381 occurred the significant but futile Peasants' Revolt. Some authorities suggest that famine at that time was aggravated by the contraction of acreage reserved for cereals. Fencing removed much land from common pasturage. Landlords, rich peasants, and the Church created large collective farms.

What were the demands in the Middle Ages? Data are hard to come by. None are available from England, but odd sources from other parts of Europe give us an inkling. It is recorded that pris-oners in Perugia in 1312 were rationed 20 ounces of bread per day, equivalent to about 1800 calories. The average working peasant consumed approximately 2000. In 1310, a reasonably well-fed Venetian seaman got 4000 calories. Cereals formed the basic diet. Bread was baked from barley or rye, sometimes oats or mixed grains, rarely wheat. Wheat, when raised, was a cash crop providing bread for the wealthy. There were also various meal-based porridges and gruels. The common beverage, ale, was fer-mented from grain. Most peasants produced the grains they con-sumed in addition to those which they produced for cash or for the landlord. Sometimes legumes (peas) supplemented the diet. Honey was a sweetener. A fair amount of fish was eaten, but little meat. It is interesting that the men of the Middle Ages consumed less meat than the old Neolithic cultures. The majority of the peas-ants were often undernourished. In those days "good years" meant more to eat and more to sell. It will be seen how this evaluation was reversed in later times.

For 2000 years, the land was exploited with only a minimal re-turn to its fertility. Now a vicious circle developed. A plot of land could produce only two cereal crops and then it had to lie fallow. An imbalance existed between arable and pasture lands owing to the necessity of producing more and more arable at the expense of pasture to provide bread for a growing population. Additionally, due to an absence of fodder crops, there was a shortage of manure for continuous arable cropping. Forage was only just beginning to be planted around Flemish towns; the practice had not spread to England. The shortage of manure was accentuated by the social structure of the times, which decreed that the villein's sheep had to be folded on the lord's demesne giving him all the manure. The additional use of manure as fuel did little to alleviate the shortage.

Data from English manor rolls show that despite everything the medieval farmer could do, yields fell off drastically. Crop yields in the ninth century are estimated to have been 2:1, that is, two units of produce for every one sown. By the thirteenth century, 3:4 was considered good. But it was no use. Yields dropped because of the continual exhausting of under-manured, overworked, unrested arable lands. Extension of arable lands came to an end by the middle of the thirteenth century. New lands failed. And even if there were new lands to be had, there was a scarcity of good working stock (oxen) which would have enabled more acreage to be ploughed. But there was no new acreage. Meanwhile the population continued to increase. Famine stalked the land.

Famine in these days was first and foremost a matter of agricultural practice, the nature of the peasant social structure, lord-serf relationships, urbanization, the labor market, oppressive taxation, the switch from subsistence to commercial farming, and the rise in population. It was not, as some would like to believe, *simply* a matter of too many people. Neolithic agriculture was able to keep pace with the population, and land could be rested long enough or new land wrested from the forest; medieval agriculture could not cope. Black Death stopped the vicious cycle by ruthlessly cropping the human population in England by an estimated 1.4 million. How many actually died may never be known, but probably one European in three died. This "gift of God" was only temporary in the face of man's staggering fertility.

Recovery began about the middle of the fifteenth century. With it the population climb started again. The growth rate of the population in Elizabethan England was remarkable. Supply could not keep up with demand. Between 1500-1620, every fourth year was a year of famine. Population pressures were especially severe in the cities. These people did not farm and had to be fed. They created a market-garden industry. Demands of the London market, for example, led outlying agricultural regions to become specialized. As regions restricted their agriculture to cabbages, cauliflowers, turnips, carrots, parsnips, or fruit, they became interdependent. Then once again there was a check on the population. In the second half of the seventeenth century, plague revisited England and together with emigration reduced the population.

These events plus the development of drainage techniques that permitted land expansion kept people fed. In the last half of the seventeenth century and beginning of the eighteenth, England not only kept people fed but actually exported corn—even in famine years!

Then the whole complexion of agriculture changed. So great was the change that the period between 1750 and 1880 became known as the period of the Agricultural Revolution. British farmers began to increase the productivity of the soil to levels beyond the wildest dreams of medieval peasants; moreover, they accomplished this miracle without any increase in manpower. More remarkable still is the fact that unlike the industrial revolution the success of the agricultural revolution was not primarily due to mechanical innovation. The eighteenth-century methods of sowing, reaping, and threshing were no more productive than in Biblical times. The revolution began with an expansion of acreage, a reorganization of holdings, and new agricultural practices. It ended successfully with the mastery of underground drainage and, in the late nineteenth century, with the application of chemistry to farming (e.g., the manufacture of superphosphates in 1843), and, lastly, with the introduction of mechanical reaping and other comparable improvements.

The greatest innovations were manuring and the more efficient use of land. At the same time, the legal and political climate changed, as did landlord and tenant relationships and the system of leases (longer), rents, and tithing. Manure, a rare commodity in medieval times, became available by abandoning the traditional medieval use of land. Instead of maintaining a permanent division of pasture and arable land, convertible farming, in which the two basic uses of land were alternated, was developed. The next great advance was sowing of fodder. This enabled land to become heavily stocked and thus to provide more manure. The use of clover spread from Italy to Holland, thence to England by 1220. Sainfoin and lucerne were also sown. By the end of the century, turnip, which had been a garden vegetable in Elizabethan times, became a regular crop. Turnips were particularly adapted for the practice of manuring because they could be eaten on the ground by sheep all winter. Turnips themselves required manuring and constant weeding and were later replaced by swedes (Swedish turnips) and mangolds, and, still later, supplemented by early clover, kale, and vetches. All of these crops had the magnificent property of being able to enrich the soil as well as indirectly produce manure. So appreciative were seventeenth-century farmers of manure that sheep were often valued more highly for it than for their wool. In Sussex, they were routinely driven from the downs to fold at night on arable land. E. Kerridge states that a breed was developed that dropped its manure only at night in the fold. All of this meant that more acreage could be cultivated.

In the second half of the seventeenth century, these innovations were accelerated by enclosure of open fields and commons so that instead of many small, scattered, inefficient, open farms there were larger enclosed farms. Basically, therefore, enclosure was a means of increasing efficiency merely by reorganizing existing resources which made for larger coordinated farms and an increase in the area under cultivation. This was helped by the rise of a yeoman class, a peasant aristocracy, capitalist peasants, who rented demesnes, acquired farms, and consolidated and enclosed them. Old style peasantry disappeared. With enclosure, crop rotation could be carried out efficiently. Two basic forms were practiced: alternate farming on light soils and "ley" farming on heavy soils. Alternate farming was a form of crop rotation: wheat, turnip, barley, clover (the Norfolk system). Ley farming was a rotation of arable and rye grass, sainfoin or clover for 2 or 3 years, followed by plowing under.

The Agricultural Revolution meant that the medieval yield of 4 to 1 of seed rose in the seventeenth century to the unprecedented ratio of 10 to 1. Agriculture dominated life. Farming became more a matter of commerce than of subsistence. Between 1732 and 1766, England exported 24,000,000 quarters of grain. In 1750, nearly one fourth of the total wheat crop was exported. Whereas in medieval times a good crop meant a good year, in the eighteenth century good crops meant low prices, and low prices brought depression. Bad years meant high profits, and thus successful attempts were made at this time to control fluctuations in the grain market by the passage of the Corn Laws. The major cause of crop failure, however, was weather.

The branch of agriculture that could not keep up with demand was forestry. With the economic revival of the seventeenth century, industry, especially the iron industry, created enormous demands for wood. In Sussex, for example, vast areas were denuded for this reason. At the same time, the growing navies and merchant fleets used unprecedented quantities of timber. Not until March 9, 1862, when the iron clads *Monitor* and *Merrimack* met in Hampton Roads and made wooden ships obsolete, did the acute need for lumber begin to decrease.

WEATHER

In tracing the rise and fall of agriculture in Britain from its origins to the end of the Agricultural Revolution the dominant theme has been the exhaustion of land. Social and political factors were powerful forces aggravating the evils of inefficient land exploitation throughout the centuries preceeding the Agricultural Revolution and providing remedies during and after the Revolution. The primary goal was always to provide food for people, and, in the process, profit. As the population expanded, the goal was not always realized; however, the failure cannot be attributed primarily to excessive human fertility. It stemmed from the finite fertility of the soil.

Important though the socioeconomic climate was, there loomed constantly over the agricultural scene the specter of devastating weather. The most critical factors affecting the yields of major cereal crops were, and are, temperature, rainfall, and snow cover. This is true of rice. It is especially true of wheat. Widespread disasters that have occurred, if not due directly to weather, were due to attacks of rust and black scab, both related to weather conditions. Despite all the ingenuity of the farmer, backed today by the most sophisticated technology, *the greatest single cause of disaster remains weather.*

From 1943 to 1972, there was no major famine in the world. In 1972, worsening weather visited droughts upon those parts of the world that could least afford it, the Middle East and India, and flooded other areas. Many populations were rescued from starvation only because they imported grain from those countries that had amassed surpluses: the United States, Canada, Australia, and Argentina. In 1973, food finally gave out in a 2,600 mile belt stretching across the southern Sahara after a drought lasting six years. Twenty-four million deaths were averted only by a massive relief effort by nations with surpluses of grain. In the same year the Soviet Union experienced such bad harvests that it had to purchase 30 million tons of grain on the world market.

Conversely, an increase in yields, especially of coarse grains (maize, barley, oats, rye, sorghum, millets) in developing countries has always been due to favorable weather, with improvement in farming techniques and the introduction of high-yielding varieties being contributing factors.

THE ROLE OF INSECTS

To what extent have insects played a significant role in the Old World in affecting the course of agriculture? To what extent have they been the agents of actual or potential famine?

It is interesting that the depredations of insects are never mentioned in the agricultural chronicles of Europe, but it is dangerous to infer from the absence of reports about insect pests that they were not a factor in agriculture. Brief mention is made of them in legal accounts, but that is all. On the other hand, in the Near East, where they were important factors, they are mentioned:

> That which the palmer-worm hath left, hath the locust eaten; and that which the locust hath left, the canker-worm eaten, and that which the canker-worm hath left hath the caterpillar eaten. . . He hath laid my vine waste and barked my fig tree; he hath made it clean bare and cast it away; the branches thereof are made white. . . . .The field is wasted, the land mourneth. . . .Be ye ashamed, O ye husbandmen; howl, O ye vine dresser, for the wheat and for the barley, because the harvest of the field is perished.

So lamented the prophet Joel (Joel 1:4-11).

The European chroniclers were not entomologists (but neither were those of arid countries), and one of the few references to insect pests is one made by Linnaeus in 1750, in which he reported:

> A single grass caterpillar was able to destroy our meadows, so that a cart-load of hay which this year is sold for 12 Talens, last year was not sold below 50. A few little nocturnal moths can cause the loveliest orchards —where neither labor nor money had been spared and which usually produce hundreds of tons of fruit—to yield now no more than 100 apples or pears.

The truly great decimations of crops, the monumental catastrophies, have been brought about, not by insects (locusts excepted) but by weather and by plant pathogens. The Bible, Aristotle, Theophrastus, Pliny, and Homer have recorded blights, blasts, mildews, smuts, and rusts. The Romans even performed certain rites during the month of April to propitiate the goddess Robigo (today identified with cereal rust diseases), who was considered the worst crop pest (*maxima segetum pestis*). In the period from 1840 to the close of the century, Europe experienced an unprecedented series of catastrophies of fungal origin. These truly plague years were ushered in by the devastating potato murrain. Nothing so destructive had ever been seen. The disease struck fields like a midsummer's frost and spread across the continent faster than cholera among men. In England, *The Gardeners' Chronicle and Agriculture Gazette* for August 23, 1845 (15), stated: "A fatal malady has broken out amongst the potato crop. On all sides we hear of the destruction. In Belgium the fields are said to have been completely desolated." Three weeks later the blight reached Dublin.

While the potato blight was gathering momentum, the odium of the vines, a powdery mildew from America, reached epidemic proportions in England, spread to France in 1848, to Portugal and Italy in 1851, and to Madeira and Germany in 1852, practically wiping out the vineyards. There was an interim period (1848-78) when an insect nearly imposed total abstinence on the people of France. An aphid, the grape phylloxera, crossed the Atlantic and within twenty-five years had destroyed nearly one third of the vineyards. No sooner had it arrived than it was followed by a mildew that attacked the French vines in 1878 and the German and Italian vines in 1882. It was during this period also that fungus leaf-spot disease put an end to coffee cultivation in Ceylon.

Between 1910 and 1913, another fungus, the agent of the Sigatoka disease of bananas, appeared in Fiji. It reached the New World in 1934, and within three years had infected tens of thousands of acres of bananas in Honduras, ruining 80 percent of the crop. Another even more virulent fungus, a root pathogen, the Panama disease, so thoroughly permeated Honduran and Guatemalan soils that by 1920 thousands of acres had been abandoned, or in the euphemistic phraseology of the trade, "retired," as totally unfit for banana cultivation.

Other more recent examples of epidemics of plant pathogens

are equally spectacular. Some of the worst, in addition to those already mentioned, were the bacterial cancer of citrus, the chestnut blight which annihilated the American Chestnut, and the Dutch Elm Disease, all in the United States. And in the case of wheat, the worst enemies are not now insects but weather and wheat rust. In 1953 and 1954, stem rust epidemics destroyed one fourth of the bread wheat and three fourths of the durum wheat in the principal spring-wheat areas of the United States.

Some idea of the magnitude of infection of cultivated crops can be gained from figures on incidence in North America. Here, around 80,000 diseases are caused by 8,000 species of fungi. Approximately 180 species of bacteria and 500 species of viruses also infect crop plants. The principal agent damaging lodgepole pine and jack pine in Canada (there is an estimated annual loss of 9.6 million cubic feet or 10 percent in Alberta alone) is the dwarf mistletoe. It causes reduction in vigor, growth, and wood quality, and sometimes mortality. The appalling indictment could be continued for many pages.

Which brings us to the second question: Is the insect becoming a greater threat to mankind's food supply as agriculture evolves? Let us search for the answer in the New World.

AGRICULTURE IN NORTH AMERICA

Once the world became committed to agriculture, North America, New Zealand, Australia, and, more recently, Russia came farthest in upsetting the basically Neolithic pattern of small plots with a village as a nucleus by mechanizing to the ultimate and placing enormous tracts of land under intensive cultivation. Only in North America, however, has the agro-industrial revolution been so fully documented from the beginning. Nowhere else do we possess as complete a picture of unfolding events enabling us to visualize how the continent appeared before and during Neolithic agriculture, during colonial development by Europeans who brought with them a sixteenth- and seventeenth-century agriculture already highly evolved, through the Agricultural Revolution, the Green Revolution, and beyond. Agriculture in the Old World had such early origins that its history from Neolithic times has had to be reconstructed largely from archeological evidence and from civil and ecclesiastical records. From North America, on the other hand, come eyewitness accounts. Furthermore, nowhere else in the world has agricultural revolution and ecological transformation moved so rapidly on so vast a scale. The changes that required centuries of evolution in Europe were, thanks to the experience gained in Europe, compressed into a few decades. Within the span of two hundred years North America was transformed from a continent of primeval forests and rolling prairies into a land of vegetable gardens, orchards, and waving grain.

From earliest colonial times, events were conscientiously recorded by clergymen and missionaries whose interests and curiosities were aroused as much by the here as by the hereafter, by literate settlers for whose diaries no item was too insignificant to escape inclusion, by amateur naturalists and historians, by concerned men of affairs—indeed by nearly everyone who could write. By 1887, when the Hatch Act was enacted, and stimulated the organization of State Agricultural Experiment Stations on a nation-wide basis, a continuing agricultural surveillance began that has been maintained throughout the land. Consequently, the history of agriculture in America offers an unrivaled case-study of the development of the relation of insects to men. In evolutionary perspective, the time span of the change has been so incredibly short as to be unmeasurable. Nonetheless, within that instant, man has proceeded with ever-increasing efficiency to dislocate the environment of native plants and the ecology, physiology, genetics, and culture of domesticated plants. Insects have responded swiftly to the change.

To have a consecutive history of these events it is necessary not only to follow prehistoric man into the New World but to be present at his first encounter with the Europeans who eventually supplanted him. One has a choice of rendezvous with the French in Canada, the Spanish in Florida and the southwest, or the English in the northeastern and central states. The French came to North America primarily as trappers and voyageurs; the Spanish, as seekers of gold. Since the English came as homesteaders, the most accurate chronicling of agricultural events comes from the northeast. Accordingly, although agriculture in the New World began in Central and South America, it serves our purpose better to focus our attention on the northeast.

When Stone Age Asians crossed over the Bering Strait about 40,000 B.C., the face of the continent was vastly different from the way it appears today and even the way it would be if man had never arrived. Great masses of ice covered the land from the North Pole south through the area of the Great Lakes. The great cold of the glacier pushed all the ecological zones as we now know them southward. Some ice-free areas were grasslands inhabited by woolly mammoths, caribou, bison, and the Arctic fox. Others were open spruce forests, the haunt of mastodons, giant beavers, and moose. The men roving these areas were hunters and gatherers, great explorers and travellers who not only dispersed to the tip of South America in a remarkably short time after their arrival in the New World (some scholars say 1,000 years) but who also pushed back ever northwards as soon as the ice began to retreat. It is a measure of their restlessness that they covered such vast territories at a time when there were no population pressures or absolute food shortages to motivate them. These earliest people, who are known to us only from the exquisite fluted blades with which they tipped their spears, hunted woolly mammoths right up to the edge of the ice. They lived within a hundred miles of the last major Wisconsin (i.e, fourth glacial stage in North America) moraine.

As the ice retreated, the ecological zones advanced northward with the gradual change in climate. Plant successions changed until, around 3000-2000 B.C., the characteristic ecological zones as we know them today became established. Simultaneously, Pleistocene animals, man's main food staple, also moved northward, only to leave the stage completely within a short time. Man sought other forms of meat and began to supplement his diet with plants. As early as 1,000 years ago, milling stones appeared for the first time. In themselves, however, they do not tell us whether the seeds that were ground were wild or domesticated. By 10,000 B.C., the fluted-blade hunters had pushed as far east as New England. Between 8000 and 1000 B.C., they were followed by other cultures of hunters and gatherers. About this time, in the more congenial climate of Mesoamerica, man turned to agriculture. Let us leave the eastern seaboard for a few hundred years and observe the development of the agricultural practices that would eventually diffuse into what was to become the thirteen original colonies.

Until 7000 B.C. in southern and central Mexico, man subsisted on wild plants. Among the earliest plants domesticated were the bottle gourd (*Lagenaria*), the chili pepper (*Capsicum annum* or

C. fructescens), summer squash or pumpkin (*Cucurbita pepo*), and small black beans. Maize appeared in camp sites (Bat Cave, New Mexico) in the southwest United States about 3500-2500 B.C. and farther south about 5000 B.C. The oldest form of corn known is popcorn. Before 7000 B.C., wild plants probably supplied most of the dietary needs; by 5000-3000 B.C., it is estimated that wild plants supplied 80 percent of the diet and domesticated plants 7 percent. By 2000 B.C., cultivated plants probably filled 20 percent of the dietary needs. The earliest evidence of cotton dates from 5000 B.C., but it was probably wild. Fully domesticated cotton has been found in archeological sites dating back to 3500 B.C.

The diffusion of agriculture from Mesoamerica was rapid and extensive. It spread into the southwestern United States, up the Mississippi and other tributary valleys, and eventually into the Great Lakes region and the northeast; but plants never became the main staple in these areas as they did in the southwest. There, where agriculture was primary, the men were the farmers; in other areas where farming was subordinated to hunting, women were the farmers. The concept and practice of agriculture was gradually introduced into the northeast between 1000 B.C. and A.D. 1 from Mexico by unknown routes. Agriculture together with pottery and mound burial characterized the culture known as Woodland (500 B.C.-A.D. 1600).

Of the Woodland Indians, the most advanced in terms of social organization, arts, technical accomplishments, etc. were the Hopewell people of the Western half of the United States (200-400 B.C.). It is surmised that their culture deteriorated into the late Woodland Culture (A.D. 200-700) because of a minor climatic shift that shortened the growing season and affected the stability of the agricultural food supply. It was their successors, the Mississippian people (A.D. 700-1500), who practiced the most extensive agriculture: It was they who spread it to New England.

It was these people whom the Europeans first encountered. De Soto, floating down the Mississippi, described their huge fields of corn, beans, squashes, pumpkins, sunflowers, and gourds. Between A.D. 1600-1760, the Hurons had at least 23,000 acres of corn under cultivation; they also grew squash, beans, and tobacco. One French missionary actually got lost in a cornfield while walking from one village to the next. Along the Wabash, fields

of crops extended for stretches as long as six miles. But the extent
of agriculture varied from place to place and tribe to tribe. The
Hurons were the great corn growers; the Chippewa (Ojibwa)
practiced only rudimentary agriculture; the Ottawa grew some
corn and wild rice and made maple sugar; the Winnebago still
stuck to hunting and fishing.

What then was the state of agriculture in North America when
the Europeans arrived? Clearly less than one half of the continent
was under cultivation. The diet of the numerous tribes reflected
the nature of the areas they occupied. In the arctic, the diet consisted of sea mammals and fish; in the subarctic, caribou, moose,
shellfish, and some berries and roots; along the northwest coast,
fish, especially salmon; in the western plateaus, fish and wild
plants; in the plains, buffalo and berries. The Indians of the middle
western prairies were hunters (buffalo) and farmers (corn, beans,
squash, wild rice). For those of the east, farm products comprised
possibly 50 percent of the diet, which was supplemented by shellfish, berries, and game. In California (which today is one of the
great agricultural states), there was no farming. The staple diet
was acorns, supplemented by insects and small game. In the Great
Basin the diet consisted of piñon nuts, wild plants, and game. The
Indians of the southwest had a diet consisting of 30-80 percent
of domesticated plants. By this time Mesoamerica had intensive
farming and irrigation.

Maize was clearly the most important domesticated plant. It was
grown from the upper Missouri River in North Dakota and the
lower St. Lawrence (47° N. Lat.) to Chile (45° S. Lat.). In pre-
Columbian times, it provided more food than all other plants
combined. It is the basis of the modern commercial corn, which
is a cross between the so-called Indian flints (second oldest eastern
variety, A.D. 1-500 and found only in the east) and dents (found
only in the prairies and related to the dents of Mexico, A.D. 1000-
1700). The second most important plants were beans: the runner
bean, *Phaseolus coccineus* (7000-5000 B.C., in Mexico); the lima
beans, *P. lunatus* (4800 B.C., in Peru); the common bean, *P. vulgaris* (5000-3000 B.C., in Mexico). One hundred and fifty-five
other species of plants were being cultivated. In addition to the
ones already mentioned, there were: edible roots (yam, potato,
sweet potato, manioc, arrowroot); edible gourd-like fruits (squash
and pumpkin); guava, papaya, pineapple, prickly pear, sapodilla,

sapota, soursap, anona, avocado, tomato, elderberry; stimulants and narcotics (cacao, coca, 16 species of tobacco); condiments (vanilla, chili pepper); fiber plants (cotton, sisal, and other Agave fibers); dye plants (indigo); resin plants (copal), hosts for useful insects (wax and cochineal insects); hedges; utensil plants; and ornamentals.

Impressive though the cornfields of the Hurons were, the extent of agriculture practiced east of the Mississippi prior to settlement by Europeans was, by modern standards, negligible. The Eastern Woods Indians, the great Algonquin stock, were half-nomadic hunters and fishermen; however, as every school child knows from reading about the Pilgrim Fathers, these Indians did cultivate corn. The earliest detailed account of their agricultural practices was written by Samuel de Champlain (6) on the occasion of a visit with Sieur de Monts in 1605 to what is now Saco, Maine.

We saw their Indian corn which they raise in gardens. Planting three or four kernels in one place, they can heap about it a quantity of earth with the shells of the signoc [horseshoe crab] before mentioned. Then three feet distant they plant as much more, and thus in succession. With this corn they put in each hill three or four Brazilian beans which are of different colors. When they grow up they interlace with the corn, which reaches to the height of from five to six feet; and keep the ground very free from weeds. We saw there many squashes and pumpkins, and tobacco which they likewise cultivate.

Thirty years later, William Wood (6) wrote in his book *New England's Prospect*:

Many wayes hath their advice and endevour beene advantagious to us; they being our first instructors for the planting of their Indian Corne, by teaching us to cull out the finest seeds, to observe the fittest season, to keepe distance for holes, and fit measure for hills, to worme it, and weede it; to prune it, and dress it as occasion shall require.

The artificiality of domestication is already apparent in the weeding, pruning, dressing, and selection of seed. "To worme it" is the earliest and only reference to pests attacking Indian agriculture that history reports.

Indian corn fields were mere scratches on the surface of the continent. In the east, the forests were still so dense and extensive that, as the saying went, a squirrel could travel from the Atlantic seaboard to the Mississippi without ever once descending to the ground. Yet the larger plantings exceeded anything that the earliest colonies essayed. A fascinating account of the extent of early colonial farming is given in Day's *History of Maine Agriculture*. This description, with local variations, was probably typical of the whole eastern seaboard. The first permanent settlers set out tiny patches of corn and small grains. Crops must have been small because time and time again one reads that the supply of bread grains was exhausted two months before the subsequent harvest. In short, there was scarcely enough to tide the families over the winter. There is no indication of crop failure, only insufficiency. One of the larger plantings was that of John Winter, who, in 1634, "paled" a field of four to five acres and set all with corn and pumpkins.

By 1666, there were European settlements scattered all along the Atlantic coast from Nova Scotia to Florida. Agriculture was family style, restricted in size and in kinds of crops. By the beginning of the eighteenth century, farming formed a thin interrupted strip in which the settlers grew, in addition to Indian corn, beans, squashes, tobacco, and an assortment of crops of European origin. Most of these were pot herbs and root crops. By this time, most of the Indians had abandoned agriculture (it was women's work anyway) and bartered pelts for food. By the end of the century, orchards of apple, pear, quince, plum, cherry, and peach were set out in many places. After the Revolution, a growing thirst for hard cider stimulated the establishment of new orchards.

About this time, the colonial farmers began to take notice of pests and other disasters that visited their crops. Drought ranked as the principal scourge of the colonial farmer. The Reverend Thomas Smith, pastor of the First Church of Falmouth, Maine, from 1727 to 1795, wrote (6) that 1736 was a "wonderful year for grass" and "the fowls and chickens destroyed the grasshoppers." In 1754:

> I have no grass growing in my mowing field, and there is no food on the Neck; the reasons are, the open winter, three weeks early drought, and the grasshoppers.

In 1756:

We are visited with the sore judgment of the worms that we were thirteen years ago, which have destroyed whole fields of English and Indian corn in diverse places.

Orchards were troubled by aphids, cankerworms, coddling moths, and tent caterpillars. A new, unknown borer also appeared. The apple fruit fly was still unknown. Apple scale probably existed, but was unrecognized and caused little damage. With the exception of black knot on plums and cherries there were no fungus diseases that caused any concern.

Elsewhere, it is recorded that in 1748 the people of New Jersey, Pennsylvania, and southern New York were compelled to abandon the cultivation of peas because of a small insect formerly rare but having multiplied excessively in the few years preceding the report. It was also noted that army worms grievously damaged meadows and cornfields. The following year, an outbreak of grasshoppers in Nahant, Massachusetts, reached such alarming proportions that the inhabitants "formed a line and with bushes drove the grasshoppers into the sea by the millions."

On the opposite shore of the still unexplored continent similar scenes had been enacted twenty years earlier and were to be repeated at intervals for another hundred years. Agriculture on the west coast was confined to the Spanish missions. Grasshopper invasions were noted by A. S. Taylor as early as 1722,

when they made their appearance and then ceased until 1746, and for three years immediately following without interruption. After this they did not return until 1753 and 1754, and finally again, before the expulsion of the fathers in 1765, 1766, and 1767.

Since 1823 the grasshoppers have several times ravaged the fields and gardens of the Franciscan Missions of upper California. About the year 1827 or 1828 they ate up nearly all the growing crops, and occasioned a great scarcity of wholesome food. At the Mission of Santa Clara, Padre Jose Viadere fired the pastures, and getting all his neophytes together made such an awful noise that those [presumably the grasshoppers] which were not killed by the smoke and fires were frightened off so thoroughly as to save the grain crops and the mission fruit gardens. About 1834-35 occurred another visitation of the grasshoppers, when they destroyed a second time the crops of the rancheros and missions with the exception of the wheat.

From the periodical press we learn, that, up to the 11th of October,

1855, and commencing about the middle of May, these insects extended themselves over a space to the earth's surface much greater than has ever before been noted. They covered the entire territories of Washington and Oregon, and every valley of the State of California, ranging from the Pacific ocean to the eastern base of the Sierra Nevada; the entire territories of Utah and New Mexico; the immense grassy prairies lying on the eastern slopes of the Rocky Mountains; the dry mountain valleys of the Republic of Mexico, and the countries of Lower California and Central America.

Col. Warren, editor of *California Farmer*, dated July 2, 1855, concerning locusts stated: "for the last three days, the very air has been so full of them over this city (Sacramento) as to resemble a dense snow storm. Large fields of oats and wheat have suffered in Ione and other upper (Sierra Nevada) valleys."

The Sacramento Union, July 2, 1855, stated: "the most remarkable circumstance we have ever been called on to notice in this locality was the flight of the grasshoppers on Saturday and yesterday. For about three hours in the middle of the day the air, at an elevation of about two hundred feet, was literally thick with them, flying in the direction of Yolo. They could be the more readily perceived by looking in the direction of the sun. Great numbers fell upon the streets on Saturday absolutely taking the city by storm and yesterday they commenced the wholesale destruction of everything green in the gardens of the neighborhood. Their flight, *en masse*, resembled a thick snowstorm, and their depredations the sweep of a scythe. The prevalence of the scourge is explained by Dr. T. M. Logan as being attributable to the great warmth and dryness of the present season—circumstances favorable to an early development of the eggs of the insect, which is deemed one of the most fruitful in the animal kingdom.

Back in the east the number of injurious insects seemed to increase with the passing years. In 1770, an outbreak of army worms spread from Langston, New Hampshire, to Northfield, Massachusetts, a distance of thirty-three miles. By 1768, pests of stored grains had become very destructive. The Angoumois grain moth, believed to have been imported in grain from France to North Carolina about 1728, spread rapidly to Virginia, Maryland, and Delaware. The home remedy was to kill it by excluding air from stored grain. This was accomplished by covering the grain carefully with layers of hay or straw. By 1774, the spring cankerworm, which had first come to notice in New England in 1666, began seriously to damage apple and plum. Again in 1794 and 1795, it was cause for concern.

In 1776, there appeared a new insect on the western end of Long Island. By 1779, it had spread to the eastern end of the island, to Shelter Island, and to Staten Island, where it severely damaged wheat crops. During the next ten years it extended its range into New Jersey and Pennsylvania. By 1789, it had reached Saratoga, New York, two hundred miles from its point of origin. It crossed the Alleghenies in 1797, reached Ohio in 1840, Michigan in 1843, Kansas in 1871. This newcomer was the Hessian Fly, introduced in some unknown manner from Europe, but probably not in the straw bedding of the Hessian troops, as believed at the time. (The Hessians were also blamed for the appearance of the chinchbug, a native insect, in North Carolina.) Two years before the bug had begun to damage crops to a noticeable degree, the British Army, with a detachment of German auxiliaries, had marched through the state en route to the battle of Guilford. A resident later remarked that it was "immediately after this event that the Hessian Fly or Hessian Bug [the chinchbug was confused with the Hessian Fly at that time] destroyed their crops of wheat; and they believed and do believe to this day (1839) that these soldiers left the flies or bugs as they passed through the country." The concept of biological warfare is clearly not a modern one! And up in Boston the bug was referred to as "the Gage bug," dubiously honoring the cordially disliked British Commander of that city.

The pattern of man's position vis-à-vis insects clearly emerges from the old records. The evolutionary relation between plants and insects was hardly ruffled by the Neolithic agriculture of the sparse populations of Indians. The advent of Euorpeans did not alter the situation at the beginning, but as the population of men increased, and the acreage and variety of crops increased, the relations between insects and plants changed.

At the beginning, the ancient cultivated crops, corn and squash, and the wild crop, hay, suffered mainly from periodic population explosions of native insects. Army worms and various species of grasshoppers, the red-legged locust, the lesser migratory locust, and the devastating grasshopper, had their lean years and fat years. Anything green that grew in their path fell to their appetites. Crops just happened to be there. A new phase began with the introduction of foreign insects. Indeed the pestiferousness of insects in colonial times was in direct proportion to the opportunities that colonists gave natural and imported insects to encounter massed stands of susceptible plants.

An early American botanist, Dr. Benjamin Smith Barton, sagely observed in 1799 (26) that, "Many of the pernicious insects of the United States seem to be increasing instead of diminshing. Some of these insects which originally confined their ravages to the native or wild vegetables have since begun their depredations upon foreign vegetables, which are often more agreeable to their palates." Not quite one hundred years later, C. V. Riley, the Chief of the Division of Entomology of the federal government, was to write (20):

It is a common remark that insect enemies are on the increase. In one sense this is undoubtedly true; *i.e.*, the number of insects affecting our fruits as well as our other crops constantly grows as our knowledge of them becomes more complete; but I question whether more injury is done today to our fruits than was done fifty years ago or a hundred years ago. . . .As the area of fruit cultivation increases so does the aggregate of injury and also the number of species we have to contend with.

Cultivation, expansion, the introduction of new plants into areas from which they had previously been absent, and the un-witting but none-the-less wholesale introduction of new insects from overseas, from one state to another as transportation became more extensive and rapid, and the increased movement of human population, all contributed to the pestiferousness of insects. The exchange of species was eminently equitable; they were trans-ported out of, as well as into, the United States. By 1912, approxi-mately 92 species of economically important insects had been im-ported. And the exchange began early, as the account of Hessian Fly and Angoumois grain moth attest. On the west coast of the United States, the importation began almost as early as the begin-ning of agriculture and has not stopped to this day.

Serious agriculture commenced in 1769 at Mission San Diego. Soon the granary weevil (*Sitophilus granarius*), the rice weevil (*S. oryzae*), and bean weevil (*Mylabris obtectus*) arrived with food and seed cereals. Later, livestock ships carried blowflies and flesh-flies (*Musca domestica, Stomoxys calcitrans, Calliphora erythrocephala, C. vomitoria, Lucilia sericata, L. caesar, Phormia regina*). Whaling, fur, hide, and tallow ships brought the red-legged ham beetle (*Necrobia rufipes*), the buffalo carpet beetle (*Anthrenus scrophu-lariae*), the hide and tallow dermestids (*D. marmoratus*), and the white-marked spider beetle (*Ptinus fur*).

The nineteenth century witnessed unprecedented dispersal of the white man throughout the North American continent. Completion of the transcontinental railroad in 1869 accelerated the process. Trade with Europe and the Far East burgeoned. In particular there was large-scale introduction of nursery stock and ornamentals, all attended by insects. Sometime around 1834, the Pear-tree Psylla was introduced. A huge outbreak occurred in Maryland in an orchard of 20,000 trees owned by Captain R. S. Emory, probably brought in on nursery stock from New York. Previous to this the insect was rare. After one year it had almost disappeared. About this time, at least not later than 1840, citrus mealy bug arrived from Europe. In 1870, the San Jose scale erupted in California, introduced, not from Chile as originally believed (it went *to* Chile from the United States) but from China, on stock consigned to James Lick. The notorious boll weevil (*Anthonomus grandis* Boheman), which had forced the abandonment of cotton in some areas of Mexico, and, with other insects, wiped out the Sea Island cotton industry on the Atlantic seaboard, was first collected in 1894. Within thirty-one years it had invaded more than 600,000 square miles of cotton, assisted in great measure by man's transportation of cotton and cottonseed. Meanwhile the citrus growers in California were struggling with a new pest, the cottony cushion scale (*Icerya purchasi*) which had been introduced in 1869 to Menlo Park, California on some *Acacia* nursery stock from Australia. Pear thrips coming from Europe about the turn of the century were reported as very destructive to prunes, cherries, peaches, almonds, and plums in the San Francisco Bay area between 1904 and 1910. The alfalfa weevil, also of European origin, was introduced into Utah in 1904. By 1910, it was a major pest. By 1917, it had spread into Idaho, Colorado, Nevada, and California.

Another major pest of cotton, the pink bollworm (*Pectinophora gossypiella* Saunders) was first described in 1842 in India. It is surmised that it spread to Egypt in shipments of seed. By 1909, it reached Hawaii and brought cotton growing to an end. It came to Mexico in 1911 via seed from Egypt. By 1917, more shipments of infected seed carried it from Mexico into Texas. Since then, collections of live moths made from airplanes at 3000 feet show that the moths are prefectly capable of dispersion by themselves over great distances. The corn borer, another moth, arrived in the vicinity of Boston in a shipment of Hungarian broom corn sometime around 1907. By 1916, it had occupied 400 square miles of surrounding countryside and by 1956 dwelt in all corn-producing areas.

Although most foreign insects arrived as hitchhikers on nursery stock, seed stock, vegetables, packing materials, etc., the gypsy moth was a startling exception. A French astronomer, Leopold Trouvelot, employed at Harvard Observatory, was conducting some genetic experiments with various silk-spinning caterpillars at his home in Medford, Massachusetts. His aim was to develop a strain of caterpillar resistant to pebrine disease, which was well on the way to destroying the French silk industry. He had purposely imported some eggs of the gypsy moth. In 1869, some of these insects accidentally escaped from Trouvelot's laboratory. He immediately notified the scientific community of the mishap, but the gypsy moth remained practically hidden for twenty years, during which time it was breeding inconspicuously but efficiently in the wastelands around Medford. In 1889, a tremendous plague of caterpillars suddenly descended on the town. The events of that year are vividly recorded by L. O. Howard in his history of entomology (10):

The numbers were so enormous that the trees were completely stripped of their leaves, the crawling caterpillars covered the sidewalks, the trunks of the shade trees, the top and the sides of the houses, entering the houses and getting into the food and into the beds. They were killed in countless numbers by the inhabitants, who swept them into piles, poured kerosene over them and set them on fire. Thousands upon thousands were crushed under the feet of pedestrians, and a pungent and filthy stench arose from their decaying bodies. The numbers were so great that in the still summer nights the sound of their excremental pellets on the ground sounded like rain. Valuable fruit and shade trees were killed in numbers by their work, and the value of real estate was very considerably reduced. So great was the nuisance that it was impossible, for example, to hang clothes upon the garden clothesline, as they would become covered with the caterpillars and stained with their excrement. Persons walking along the streets would become covered with caterpillars spinning down from the trees. To read the testimony of the older inhabitants of the town, which was collected and published by a committee, reminds one vividly of one of the plagues of Egypt as described in the Bible.

During all this time the Medford people had been under the impression that the insect which they were fighting in their gardens was a native species, and they knew it simply as "the caterpillar" or "the army worm"; but in June, 1889, when the plague was at its height, specimens were sent to the Agricultural Experiment Station at Amherst, and identified by Mrs. C. H. Fernald as the famous gypsy moth of Europe.

A town meeting was immediately called in Medford, and work against the insect was begun.

Dispersal was by no means a one-way flow, from the Old World to the New. New World insects were finding their way across both oceans. The Colorado Potato Bettle is only one of many that emigrated from America and became pests in their new environment. The danger from unintentional international trafficking in insects became all too apparent at an early date, and efforts at quarantine were made. On February 5, 1898, for example, Wilhelm II, Emperor of Germany, alarmed at reports of the destructiveness of the San Jose scale in California, and its spread eastward, issued a decree prohibiting the admission of American fruits and living plants into Germany. This decree was the stimulus for the establishment of international quarantine on a wide scale. Austria-Hungary followed with a similar decree. The political repercussions engendered by these decrees were as great as they were ridiculous. The quarantine was declared to be a retaliatory move by Germany for certain unfavorable tariff legislation enacted by the United States!

By now, however, the damage had been done. By his agricultural practices, man had set the stage for native insects, the migratory grasshoppers, army worms, chinch bug, Colorado potato bettle, grape phylloxera, plum curculio, corn earworm, and others, to multiply on the abundance of food provided and to follow the crops into new areas as these were opened to agriculture by clearings, irrigation, and cultivation in general. By his commerce he had introduced foreign insects, the cotton boll weevil, Mexican bean weevil, codling moth, European corn borer, Hessian fly, Japanese beetle, alfalfa weevil, citrus scales, and numerous others, to greener pastures. Any lingering doubts that one may have entertained about the reality of an increasing "insect problem" are dissipated by these accounts of the European's first century in the New World.

The appearances and ascendancy of insect pests are reflections of man's advances in colonization, agriculture, interest in new kinds of crops, monocultures, continuing interference with ancient insect-plant relationships, and as we shall see later, politics and economics. Nevertheless, granted that there are new insects and that many native insects have transferred their attentions to cultivated plants, the evidence that insects compete *seriously* with us for *food* is unconvincing. Weather, plant pathogens, and complex socioeconomic factors are the principal agents that threaten our food supply. The role of the insect is indeed becoming more dominant as agriculture evolves, but criteria other than famine must be employed to assess it. The black mount of the Third Horseman of the Apocalypse, Famine, rides in many guises, but he is not an arthropod.

References

1. Anon. Review of the World Grains Situation. Internat. Wheat Council, London, 1968, 79 pp.
2. Chambers, J. D., and Mingay, G. E., *The Agricultural Revolution 1750-1880*. B. T. Batsford, London, 1966, 222 pp.
3. Clark, A. H., The impact of exotic invasion on the remaining New World mid-latitude grasslands. In: *Man's Role in Changing the Face of the Earth*, Vols. 1 and 2 (W. L. Thomas, ed.). University of Chicago Press, Chicago, 1956, pp. 737-62.
4. Curtis, J. T., The modification of mid-latitude grasslands and forests by man. Ibid., 1956, pp. 721-62.
5. Darby, M. C., The clearing of the woodland of Europe. Ibid., 1956, pp. 183-216.
6. Day, C. A., History of Maine Agriculture 1604-1860. Bul. Univ. Maine 56 (11): Univ. Press, Orono, Me., 1954, pp. 1-318.
7. Driver, M. E., *Indians of North America*. 2nd revised ed., University of Chicago Press, Chicago, 1969, 632 pp.
8. Duby, G., Medieval Agriculture 900-1500. In: *The Fontana Economic History of Europe. The Middle Ages.* (C. M. Cipolla, ed.). Collins/Fontana, London and Glasgow, 1969, pp. 175-220.
9. Essig, E. O., *A History of Entomology*. Macmillan, New York, 1931, 1029 pp.
10. Howard, L. O., *A History of Applied Entomology*. Smithsonian Misc. Pub. Washington, 84 (1930), 1-564.
11. Irving, W. N., and Harrington, C. R., Upper pleistocene radiocarbon-dated artefacts from the northern Yukon. Science, 179 (1973), 335-40.
12. Jennings, J. D., and Norbeck, E. (eds.), *Prehistoric Man in the New World*. University of Chicago Press, Chicago, 1964, 633 pp.

13. Kerridge, E., The sheepfold in Wiltshire and the floating of the water meadows. Econ. Hist. Rev., 2nd Ser., VI. (1953-1954) pp. 282-86.

14. Kramer, S. N., *The Sumerians*. University of Chicago Press, Chicago, 1963, 335 pp.

15. Large, D.E., *The Advance of the Fungi*. Jonathan Cape, London, 1940, 488 pp.

16. Malenbaum, W., *The World Wheat Economy 1885-1939*. Harvard University Press, Cambridge, 1953, 262 pp.

17. Martin, P. S., The discovery of America. Science, 179 (1973), 969-74.

18. Narr, K. J., Early food-producing populations. In: *Man's Role in Changing the Face of the Earth*. Vols. 1 and 2 (W. T. Thomas, ed.). University of Chicago Press, Chicago, 1956, pp. 134-51.

19. Quimby, S. I., *Indian Life in the Upper Great Lakes. 11,000 B.C. to A.D. 1800*. University of Chicago Press, Chicago, 1960, 182 pp.

20. Riley, C. V., *Insect Life*, Vols. 1-7, 1888-1895, Washington, D.C.

21. Roehl, R., Patterns and structure of demand 1000-1500. In: *The Fontana Economic History of Europe. The Middle Ages.* (C. M. Cipolla, ed.). Collins/Fontana, London and Glascow, 1970, pp. 105-142.

22. Smith, E. C., and Stephens, S. G., Critical identification of Mexican archaeological cotton remains. Economic Botany, 25 (1971), 160-68.

23. Taylor, A. S. (1859), An account of the grasshoppers and locusts of America. Smithson. Inst. Ann. Rpt., 1848, pp. 200-213.

24. Underhill, R. M., *Red Man's America*. Revised ed. University of Chicago Press, Chicago, 1971, 395 pp.

25. Usinger, R. L., The role of Linnaeus in the advancement of entomology. Ann. Rev. Ent., 9 (64), 1-16.

26. Weiss, H. B., *The Pioneer Century of American Entomology*. Published by H. B. Weiss, New Brunswick, N. J., 1936, 320 pp.

Chapter Three
Starvation Versus Profit

3
Starvation Versus Profit

In the preceding chapters we have seen how our present relationships with insects have originated and developed and to what extent insects are competitors for plant material. It is now appropriate to scrutinize the magnitude of this competition more closely and to attempt some quantitative assessment of its impact. Only then can we determine what our attitude toward insects should be —how they should be treated by mankind, whether or not they should be annihilated, and how great a price in adverse side effects we are willing to pay. Miscalculations can create disturbances that reverberate in every sphere of our existence; therefore, it is imperative that relationships be properly understood and dispassionately analyzed.

GAINING A PERSPECTIVE

As many authorities and august bodies have repeatedly pronounced, the food problem is one of the world's most acute and pressing problems; it is responsible for much global unrest, and, with each succeeding generation, it becomes more acute. Who can refute this assertion with the knowledge that people are dying every day from starvation or suffering from chronic malnutrition? Even in this most advanced technological year, 1975, there is famine across the breadth of Africa. There is widespread malnutrition in the world. In 1973, more than 300 million children suffered retarded physical development, and perhaps even stunted mental and behavioral development, for lack of sufficient protein. While part of the world dines on filet mignon and caviar or diets to reduce obesity, even more of the world moans despairingly for lack of calories. The disparate distribution of the world's food supply is fully documented in an editorial in the January 24, 1975, issue of *Science* and in the Foreign Agricultural Economic Report of December 1974. On a global basis there simply is not enough food, even if it were evenly distributed, to keep the present population in the best of health. From this, it follows that we can ill afford to lose or waste any food. Furthermore, as long as man continues to be almost totally dependent on three grains—maize in the New World; wheat in Europe, the Near East, temperate Asia, and North America; and rice in the Far East—massive failure of any one of these crops can lead to famine. Failure can also deny man his animal protein because his cows, chickens, and other domestic animals are sustained by cultivated cereals. Of these facts there is no doubt.

From the presence of malnutrition and periodic famine, together with the continuing rise in population, many people have argued that damage to crops by insects cannot be tolerated and that failure to prevent this damage means starvation. Consider some of the following statements.

A Nobel laureate, Dr. Norman E. Borlaug, was recently quoted by the Philadelphia *Bulletin* as saying: "It is as simple a matter as that. We can either use pesticides and fertilizers at our disposal or starve. Deny them [agricultural chemicals] to us and the world's population will expand at a rate much faster than we can produce food." At the same time, this authority remarked that "the efficiency of our farmers in the United States is so great that one man can now produce enough food for forty people. In great part this comes from his larger use of chemical fertilizer and pesticides." The fact that fewer farmers on smaller acreage produce more food than ever before argues for greater efficiency but does not prove that the alternative to efficiency is famine. History has shown that the argument for fertilizer, whether it be manure or chemical, is valid. However, the argument that losses inflicted by pests cannot be borne because the cost is starvation must be examined more carefully.

In the *New York Times* of July 11, 1971, an official of a leading chemical company is quoted as believing that the United States could not produce enough food without pesticides. He estimated that withdrawal of pesticides from agricultural production would reduce the total output of crops and livestock by 30 percent. He also asserted that the elimination of pesticides would increase the price of farm products by 50 to 75 percent.

These references, and they could be multiplied many times, are excerpted from daily newspapers rather than from professional, and therefore presumably more reliable, journals because it is precisely these reports that are read by the man-on-the-street and help shape public opinion. Many official reports, and those designed to shape political opinion, do, however, take the same approach even though in a more restrained tone. A lead article in the August 24 (1973) issue of *Science* begins with the sentence "Reports of famine in Africa and India, depleted grain resources in the United States, and the current soaring food prices have all emphasized the necessity of controlling insect pests that are man's major competitors for food and fiber." An authoritative book on insecticides contains the statement: "Modern pesticides at the moment are essential to man's existence." A report of the National Academy of Science in Washington states that: "Considered from the standpoint of the world situation and our own commitments to underfed nations our present food and feed surpluses are in fact little more than prudent insurance against a series of unfavorable crop seasons."

An official brief (12) published for the Entomological Society of Canada states:

The current demand for food is great nationally and (despite local 'surpluses') is very great internationally. This demand is growing rapidly because of increasing numbers of people and rising incomes. Pests exact a heavy toll of man's crops: An authoritative estimate by the Food and Agricultural Organization of the United Nations in 1969 put the average loss due to pests (not including diseases or weeds) at 12 percent with a range of 5-20 percent. Regardless of the steps that may be taken to reduce or stabilize human numbers and human demand for food, it is clear that even now we can ill afford the loss that pests would inflict were they not controlled.

At the same time, Glenn T. Seaborg, in his address as the retiring president of the American Association for the Advancement of Science, pointed out that: "It is an appalling fact that, at this time of increasing food production, as much as a quarter of the world's food never reaches the consumer because it is destroyed *in storage, in transit,* or *in the market* [the italics are mine] by insect pests, rodents, bacteria, and mold. In some areas the loss runs as high as 80 percent of the supply produced." There is further documentation in an article in the *Christian Science Monitor* (October 16, 1974) headlined "Bangladesh: Famine has many causes." Among the causes mentioned are corruption, smuggling, hoarding, and a poor distribution system. A further reference to the Canadian brief is illuminating:

There are probably several crops which could sustain pest damage greater than is now tolerated without sacrifice of nutritional quality; but at the same time there is no doubt that much of the agricultural produce in Canada would be inedible or unmarketable unless protected from pest attack. Where forest pests and biting flies or nuisance insects are concerned, priority must be placed on different criteria but the main conclusion would be the same: For some resources, effective pest control will remain essential if standards of human health and comfort [and, one might add, profit] are to be maintained at, or even somewhat below, present levels.

The theme of starvation, explicit or implicit, recurs in report after report. That pests can tip the balance toward or away from starvation is difficult to accept at face value because of the great impact of socioeconomic factors on crop production and food availability and because of the crudity of methods for arriving at estimates. There is reason to believe that plant protection specialists tend to overestimate losses. The specialists feel that they underestimate. The theme is still further suspect when one considers that local losses exceeding those occasioned by pests are periodically and deliberately incurred by activities aimed at maintaining high prices. Think only of the mountains of coffee burned, potatoes left to rot, crops plowed under and subsidies paid for *not* planting crops. Consider the fact that at one point cotton supply outran demand with the result that acreage was restricted by national policy and price support was instituted. Between 1940 and 1965, every cotton-growing state except Arizona, California, and New Mexico reduced acreage. Think of all the subsidies and laws applied to agriculture, from the eighteenth-century Corn Laws of England to the present, to regulate the prices of produce. For a modern example, consider what happened to palm oil in Nigeria in 1964: In order to provide easy revenue, the Nigerian government sold the oil abroad at a high world price while farmers received only half that price. Naturally production declined. Palm oil became so cheap locally that it was wasted by being used for fuel as a substitute for kerosene.

Consider further the fact that some parts of the world produce huge surpluses while other parts starve. As the National Academy of Sciences reported, the United States produces more food than it needs and "suffer(s) not from a food shortage but from food surpluses, which are often embarrassing to us politically, on both the domestic and political fronts." It was pointed out at the same time "that there is a world food problem of almost overwhelming proportions. In India, for example, the food supply is so precarious, that when drought, flood, or early frost stike any part of the country, famine and mass starvation can be avoided only if the central government moves in quickly, often with grain provided from the United States from its current surpluses." Note again the reference to climatic factors!

The fact that parts of the world have surpluses while others starve does not reflect a pest problem. Clearly, weather, agricultural practices, economic factors, and the stages in the development of modern agriculture that different countries have attained underlie the discrepancy. The principal factors responsible for American surpluses have been mechanization, improved varieties, increased use of fertilizers and pesticides, and several years of unusually favorable weather. We do not control pests any better than the rest of the world.

APPLES AND ORANGES

What then actually determines the amount of food made available by agriculture? What accounts for surpluses? What accounts for deficits? Let us examine the business of food production first in highly industrialized nations. In the process we may discover to what extent insects affect production and supply. Basic to this enquiry is an understanding of the relations between production, supply, and starvation. The complexity of the economics of agriculture is seldom appreciated by entomologists specializing in pest control. As a result they tend to oversimplify the relation between insects and agriculture.

Production is not directly related to nutritional need; it is related to *price*. Let us, therefore, consider price for a moment. Farmers are price takers. Like everybody else, they want to make a profit. Repeatedly the maxim is stated that every other consideration must be subordinated in favor of the cultivation of the crop that will bring in the largest profit. The stimulus to which production responds is price! If price were merely a matter of demand, there might be some relation between need and supply; however, as we shall see, prices are manipulated. Furthermore, demand itself is a very complex variable; it is not just a matter of how many bellies are to be fed. Let us, therefore, consider price and demand.

In purely competitive, unregulated markets there is a free interplay of supply and demand. The following example is taken from a book on the economics of agriculture by the British economist Capstick. Consider a quick-growing crop like lettuce. A farmer brings his first crop to market and decides that the price he is receiving is lower than normal, whereupon he concludes that the market is oversupplied; consequently, he reduces his next sowing. If other growers reached the same conclusion, they too reduce their sowing. If demand remains steady and supply is adequate, price is stable. If there is instead an unsatisfied demand, there will be a rise in price, growers will sow more lettuce, price will fall, the next round of sowing will be smaller, and so on. But the relations between supply and price are complicated by many factors, including, among others, the perishability of the product, the vagaries of wholesalers, and actions of price speculators. Furthermore, few markets are purely competitive. Because physical conditions of uncertainty characterize farming and because farmers are price takers, governments and other agencies attempt to introduce an element of certainty into marketing. Their measures have an enormous influence on supply. They include not only control of prices but also control of production. Both make it possible to guarantee price, and a guaranteed price will stimulate production.

There are a number of reasons why governments introduce guaranteed prices. One is insurance. It may be desirable to keep agriculture going on at a higher level than necessary in order to counterbalance potential war, transportation or labor crises, or undesirable trends in balance of payments. It may be desirable to stabilize the incomes of farmers in order to protect the national economy. Also, it may be desirable to raise the income of farmers so that there will not be great disparities between the agricultural and industrial sectors. Otherwise, how do you keep the boys down on the farm? Even in developing countries, filling hungry mouths is only one small factor. David E. Bell of the Agency for International Development has stated the situation clearly: "So it would be a mistake, I think, to conclude that the only reason for placing high emphasis on agricultural development in developing countries is in order to meet food supply problems. Economists do not hold that to be the case. The reason for placing high emphasis on food production, where the resources permit, relates to the normal economic criteria of maximum return on investments and other inputs."

Another way to control supply is to store the surplus produce of the fat years for release in the lean years. Still another is to impose quotas. Control of supply is also achieved by withdrawing land from production, paying farmers not to plant so many acres of crop. In this last case the farmer usually withdraws his poorest land from production so the effect on supply is in fact negligible. When a crop is very profitable, farmers tend to increase cultivation of that crop at the expense of others. To counteract this tendency, governmental strictures are applied. In England, the Potato Marketing Board registers all potato growers and restricts them to fixed acreages. Cropping restrictions on beets are imposed by the British Sugar Corporation, which literally owns all sugar beet seed. On the international level, all sorts of agreements covering the marketing of agricultural commodities are in force. These lead to situations where crops or products must be deliberately destroyed. At one point, for example, Brazil was destroying its coffee trees because of overproduction.

The United States bolsters agriculture by support buying when prices fall below parity (the relation between prices paid by and received by farmers based on the years 1910-1914 as a norm) and by paying farmers to withdraw land. Despite withdrawal of 20 million hectares from production (nearly twice the agricultural area of England and Wales) and less than half the peak number of farms and farmers, supply continues to increase. It has been estimated that by 1980, United States Agriculture could produce surpluses with as few as 700,000 farms instead of the present three million, and with a labor force of 2.5 million instead of the present 6 million.

The demand for products is based upon the whims of millions of consumers. As population grows, demand increases; but demand on a global basis is uneven. How demand will actually relate to production depends on the intricate patterns of world trade. What and how much a farmer will plant to anticipate this long-term demand requires more accurate prophetic powers than are presently granted to most of us. This clouded vision is one of many reasons why much of the talk about crop protection and famine is nonsense. Farming is not a simple business. Additionally, as personal incomes rise, people do not buy more food (once they have satisfied their appetites), because the stomach has, after all, a finite capacity; they buy *different* foods. This is also true of developing countries, so that in this respect the demand for food will continue to increase even after starvation and nutritional deficits have been corrected and even if the population remains static.

The demand that really counts for the farmer is next year's, not the next generation's. This depends on a number of factors of which population per capita income is most important; but population projections are significantly unreliable, and per capita figures involve complexities that make predictions of demand tricky. As the income rises, the *proportion* of the increment spent on food decreases. Although the rich spend more on food than the poor, the money goes not for increased quantity but for different type and quality. Also, habit and tradition play a very great role in determining demand. It is interesting, for example, that in the United States there was a great decrease in the consumption of butter when high cholesterol in the blood was linked to heart trouble but that in Europe, where few people believed that cholesterol was dangerous, no such decrease occurred. In the short term, that is, when the farmer actually brings his produce to market, demand may also depend on season, weather, price, fashion, and so forth.

More than 50 percent of the world's population is engaged in agriculture or earns a livelihood from ancillary businesses. In tropical and developing countries, the figure is 90 percent. In Belgium, the Netherlands, the United States, and Great Britain, the figure is less than 10 percent. In the many countries that have no real division of labor, hence no market economy, each family or tribe produces its own food and all the other necessities of life. This subsistence farming is still to be found in relict areas in even highly industrialized countries. Until tourism became a business in the twentieth century, the Aran Islands off the coast of Ireland practiced subsistence agriculture. The advent of the trader (and/or tourist) turns the subsistence farmer into a commercial farmer with small holdings. And although some countries have only subsistence farmers (technically labeled peasant farmers), they can still place national agriculture on a commercial basis. Burma and Thailand, with only peasant farmers, are the world's major exporters of rice.

Thus, the kinds of farms in the world today are: peasant farms in the tropics and developing countries; family farms in the developed countries of the West and Australia and New Zealand; agrobusinesses (huge enterprises) primarily in the United States; plantations (e.g., rubber) mostly in the tropics; and state or collective farming in Communist and Socialist countries and Israel. With the exception of peasant farming, these are all business enterprises, whether privately, collectively, or state owned. Plantation agriculture, begun in the heyday of seventeenth-century colonization, is big business. Family farming is defined by the United States Department of Agriculture as "a risk-taking business in which the operating family takes most of the risks and does most of the work (like shopkeepers and hoteliers)." In the United States, a family farm provides on the average work for only three people; nevertheless, the family farm accounts for 75 percent of farms in the United States and 50 percent in the United Kingdom.

What we have discussed so far indicates that the profit motive is an especially powerful overriding factor in the relation of supply to demand. This is especially true in the case of universal staple crops. Consider the case of wheat. Leavened bread is preferred over all cereals. The only grains that can be used in the production of leavened bread are wheat and rye. Rye has almost disappeared as a staple. Wheat, on the other hand, is grown in every part of the world except the hot low-lying regions of the tropics. It has been called the most important commodity in world trade. The use of the word "commodity" rather than "food" or its equivalent is significant.

In 1951, the estimated f.o.b. value of world wheat approximated 1,950 million dollars. The employment of merchant marine fleets, ship-building industries, port facilities, and countless subsidiary activities depends to a major degree on the transportation and handling of wheat. Wheat is a peculiar commodity (as are other important agricultural crops) in that a rapid increase in price relative to that of other products stimulates increased production quickly, whereas a drop in prices does not alone cause a corresponding downward adjustment. When wheat commands high prices, production is extended into marginal lands; when prices are exceptionally high, there is expansion into prime land at the expense of other crops. The wheat farmer strives to maintain his gross income as high as possible by maximizing production in periods of declining or low prices to compensate for heavy capital overhead. When prices are very high, freight rates for wheat exceed those for oil; then tankers, bulk carriers up to 60,000 tons, clean their bunkers and enter the wheat trade.

In short, wheat is money, wheat is national economy, wheat is balance of payments on the international level. Banks, economies, governments, and nations can rise or fall on the surpluses or deficits in wheat and the price per bushel. A crop failure in wheat, therefore, at least in the developed countries, is not raising first and foremost the specter of starvation; it is raising the specter of bankruptcy. In the past, farming was essentially a self-contained enterprise. As it becomes more dependent on inputs from products produced off the farm (machinery, chemicals, etc.), money income and the stability of money income becomes increasingly important. Costs have risen faster than profits, so the farmer is forced to try to reduce costs (by substituting machinery for labor) and to protect his investment with insecticides. The 169 percent increase in pesticide sales in America between 1952 and 1968 arose as a part of the development of a highly specialized and technical agriculture. Research and development on pest control has not, for obvious economic reasons, reduced the problem.

But global statistics are remote; they do not conjure up any picture of familiarity or reality in the mind of the average citizen. Let us focus instead on a less universal staple grown in a smaller area. The total arable land in the county of Arooskegan was planted with potatoes. In the space of a few years, the Colorado potato beetle and flea beetle became increasingly resistant to chemical treatment. Potato growing became increasingly costly, then economically marginal, and finally went out of production. With the harvesting of the final crop, the farm workers were discharged. Since many of these were migrant workers, the end of potato growing merely meant that the next year they would migrate to a different area to field a different crop. As soon as the last potato of the summer had been processed, the processing plant closed permanently, throwing both the blue collar and white collar workers out of work. With no prospects for the future, the owners could not meet mortgage payments and the bank foreclosed. It now had on its hands a piece of property from which it could realize no income. Shares fell off. The single rail spur to the main town, although it had only three freights weekly, had its service reduced first to one and finally to none. The station master retired early; the freight clerk and trainmen sought employment elsewhere. The local trucking concern went out of business because the owner had operated local service from the railhead to outlying areas. He too could not meet his mortgage payments, insurance premiums, and road taxes. The local sales office for farm machinery closed. One of the three local garages that used to service trucks and tractors let two of its mechanics go, and after struggling along until January closed shop. The owner got a job as an electrician in a naval station forty-seven miles away.

As the rate of unemployment rose, some people went on relief and others emigrated. Attendance at the cinema dropped. It tried shifting from movies to folk music bands, then closed. The bank owned that building too. As the town's population gradually shrank many shops were affected. The local barber had fewer heads of hair to cut, the branch of the supermarket moved out of town, one of the two dentists left, the post office reverted to the status of a substation, the Greyhound and Trailways buses were rerouted. The age profile of the town changed. As those of reproductive age left, the population became older and hence non-productive in more ways than one. The tax base changed. Schools and tax-supported public services deteriorated. The selectmen tried to attract new industry to the town, but although they offered tax advantages, transportation links were practically non-existent, and local manpower was unskilled or elderly. The area's principal raison d'être, the potato, was gone. Nobody starved, the state as a whole was not brought closer to famine. Nonetheless, the potato beetle and the flea beetle caused the death of a town. Of course, they were not unaided. They had had the assistance of an economy based on a single crop and of a control program that sent costs soaring before even it failed completely.

Arooskegan is a mythical county, but it can be matched in real life by any county in a region whose economy is based on a single crop. Similar disasters have in fact occurred in the Lower Rio Grande Valley and in parts of Mexico where cotton growing declined from 700,000 acres to 1,200 acres in 1970 as the result of an insect pest, aided in its destructive endeavors by man and his chemical control system.

To see the other side of the coin we can take a true example, Fargo, North Dakota. The bumper crop of wheat in the early 1970s together with the bumper prices of food started a boom. In twelve months, the price of farm land increased thirteen percent. The Fargo National Bank, already sporting a six-lane drive-in-teller operation, erected a new ten-story headquarters. Fargo people bought more appliances and television sets, lake homes in Minnesota, and bigger cars. One farmer installed a mobile telephone in his tractor to handle business while plowing his fields. A new enormous shopping mall was built on the prairie; the 300-room Holiday Inn outside Fargo is nearly always booked solidly. Money is freer than it has ever been.

FARMING AS BUSINESS

Now we can begin to discern some of the true reasons for wishing to control insects. Farming is "big business." Its principal product is money; food is a by-product. To argue the need for crop protection against pests in terms of starvation is to emotionalize and falsify a problem that is anything but simple. Holding the threat of starvation over the heads of people to encourage the use of pesticides is a scare tactic reminiscent of the theological threat of Hell and eternal damnation designed to keep the faithful in moral line. It succeeds for a while, in both cases, but is unconscionable and in the end self-defeating. Furthermore, insofar as agriculture in industrialized countries is concerned, it is not supported by the facts.

All well and good you might say. In countries with developed market economies, the countries that the Food and Agriculture Organization (FAO) of the United Nations has classified as Zone A (the United States, Canada, and Western Europe), famine is a scourge of the past and pests do not raise the specter of its return. Countries in Zone B, countries with planned economies (the Soviet Union and Eastern Europe), are also in a fairly safe state. These countries represent, however, only a fraction of the world's population.

The majority of people live in what FAO calls Zone C (Latin America, Asia, the Far East, Africa), the developing countries. In these countries the *rate* of population growth is highest, the production of food per capita is lowest, malnutrition is commonplace, and recurrent famines are facts of life. The situation gets worse rather than better. There is a growing gap between food supply and need. At the present time, many of these countries rely in times of dire emergency on the charity of the developed nations. To remove themselves from this situation they have only two choices: to grow sufficient food for their own needs or to purchase from those nations that produce surpluses. There is little hope for the latter because developed nations must charge high prices just to break even let alone make a profit, and the developing nations cannot get a sufficiently high price for their raw materials to raise the necessary capital. The solution lies in increasing agricultural production or reducing the population growth, or both. Our concern here is with agriculture.

What are the principal obstacles to developing countries producing the food they require? First, there are deficiencies in the agricultural infrastructure and social provisions for agriculture. There is poor marketing organization, inadequate transportation, storage, and handling facilities, and poor education. Second, there are deficiencies in the agricultural and farm structure, many of them reminiscent of the troubles that plagued the medieval farmer. Most of the farms are peasant farms; the land is abused; there is a lack of machinery (the equivalent of medieval oxen) to expand acreage; the use of land is inefficient. Third, there is a lack of capital. Fourth, there are inadequate provisions for financial assistance: Credit is dear. Fifth, there is instability of prices.

It is argued that the solution to the problem of feeding the world lies in increasing the productivity of existing arable land rather than in bringing more land into production. In some countries, of which the United States is an example, the thrust of the argument against bringing more land into production is economic; the cost of cultivating new land is considered too high to be borne. It is also argued in some quarters that the arable land not being used is of marginal quality anyway. That assessment is false. Rich agricultural land in the United States is disappearing at the rate of several million acres per year to provide for highways, housing developments, and industrial complexes.

To increase productivity of existing land in developing countries the requirements are high-yielding varieties of crop plants, fertilizer, machinery, capital, and a restructuring of agricultural and marketing practices. Freedom from starvation hinges upon these benefits, but J. B. Knapp of the World Bank Group observed: "Simple as the notion of growing more food may appear, actual success in doing so depends on a host of interactions being taken at different levels and in different sectors of the economy."

In short, solutions to the problems that developing countries face in providing more food for their people are not merely matters of providing fertilizer and killing insects. Consider, for example, capital. Many countries cannot effectively absorb more capital now even if the World Bank provided it. Other things frequently must come first. The situation in Kenya is a case in point. Prior to 1950, agriculture was carried out on a tribal basis. Holdings were scattered and temporary and the farmers had no incentive to cherish or improve the soil. The government initiated a program of reform in which farms were consolidated, titles were registered, farmers were educated in new techniques, water supplies were developed, feeder roads constructed, and market outlets provided. Then, and only then, did the World Bank provide a loan of $5.6 million and the Agency for International Development, $4.5 million. Clearly, much agrarian, administrative, and organizationl reform is necessary before capital can be of any assistance. The picture is startlingly reminiscent of medieval times in Europe.

Even if providing more fertilizer were the only solution to increasing productivity, realizing that end is no longer as simple as it was in times past. Most fertilizers are now chemically manufactured. Making more should be elementary. The technology exists. In fact, however, it is more difficult to produce effective distribution systems than it is to produce fertilizer. Marketing is still one of the major weaknesses in the system for supplying fertilizer.

Other deficiencies are the lack of incentives for farmers, the use of cheap food policies to check inflation, too much concern about foreign exchange, ill-advised uses of export taxes on farm products, short-sighted cropping patterns that already show burdensome surpluses in some tropical countries, and unbalanced kinds of economic growth. Where there is a monolithic pursuit of industrialization, the economy is unbalanced, and agriculture suffers. Only where there is balanced economic growth, as in Israel, Taiwan, and Mexico, can the problem of supplying more food be met successfully.

Herculean efforts are being made by many developed countries to remedy the numerous defects—social, political, economic, and biological—that stand in the way of feeding the world. The United States has assumed a leading role in this respect. Although it and other nations with surpluses provide food in times of crisis, its long-range policy is not to feed the world but to help the world feed itself. Biological steps already taken in the direction of increasing yields on existing acreage have resulted in the "Green Revolution." High-yielding varieties of rice and wheat were developed in agricultural research centers in the Philippines and Mexico and immediately produced bumper crops in the field. There have also been spectacular increases in yields of beans, millets, sorghums, and maize.

But success has been followed by a rude awakening and hurt surprise. What *is* surprising is that there has been amazement at the troubles engendered by the Green Revolution. It is incredible that the troubles were not anticipated and neutralized. For anyone who had done his homework or who had knowledge of the history of agriculture in North America, the signs were as clear as the nose on his face. History was repeating itself. Better varieties selected for yield and not resistance, improved water management, richer fertilizers, larger monocultures, new cropping conditions creating favorable microenvironments, etc., etc. have favored the development of pests where there were no pests before. There are numerous reports of insect pests becoming more serious after the release of a new variety of crop.

At a conference on pest management, held at North Carolina State University in March 1970 (20), an entomologist from Pakistan focused squarely on the problem:

This Green Revolution will have a definite effect on pests and pest management. This country and its people have been very active in bringing about the Green Revoution, and we have today become self-sufficient in grains. However, I must warn you that this Green Revolution may be short-lived because of the changing complex of corresponding changes needed in pest management. For example, our wheat crop used to be free of insect pests, but with the introduction of Hessian fly resistant varieties, the species complex changed, and now there are several species of insects which are damaging wheat crops. An impression has been created here and abroad, particularly in Washington, that because the Green Revolution has been achieved perhaps further attention is no longer required to continue or to intensify production of food in less-developed countries.

FAO made the following statement in 1969: "As the constraints imposed by poor water control, low soil, fertility, and lack of responsive varieties are removed, pests, diseases, and weeds may well represent the main obstacle to the continued expansion of food production in much of Asia, the Near East, North Africa, and coastal Western Latin America."

In developing countries, the farmer does just what his counterpart did, and still does, in developed countries: He sprays insecticide copiously and haphazardly. A. S. K. Ghours of the United States Department of Agriculture remarked (20), "these areas are under great pressure from the producers of pesticides. They tried to sell us all sorts of pesticides; we are under pressure from our governments and from our farmers to use pesticides."

There is no doubt that pests are more of a problem in developing countries now than in the West. The situation is now as it was in the West one hundred years ago when there were also famines. But of all the factors responsible for famine, pests must rank low on the list. Climate and soil fertility still rank highest.

At the same time, the new strains of crops are very demanding. They require advanced technologies which farmers in developing countries are unable to learn any more rapidly than farmers in highly developed countries, and which are beyond their financial means to apply. Irrigation, fertilizer, new skills are necessary. Land supporting new strains becomes so valuable that many farmers are no longer able to afford it, and tenant farmers are evicted by landlords. Many of the socioeconomic constraints on agriculture that we have already mentioned have slowed down the Green Revolution.

Clearly, in the complex agro-socioeconomic web of the twentieth century, insects are a force to be reckoned with, but the reckoning must be rational, not an emotional appeal referring to famine. And therein lies the difficulty. As Chiarappa, Chiang, and Smith have asserted in a very illuminating article on the assessment of crop losses:

We intend to stress the fact that little reliable information on the magnitude of crop losses caused by pests and diseases is available, and that the deficiency of such knowledge is particularly acute at the farm level. Although some of this ignorance is based on the intrinsic difficulty of measuring crop losses and the lack of suitable methods for assessing them, much of it also results from lack of interest and past neglect on the part of plant protection scientists.

Analyzing losses may at first glance appear quite simple. All one has to know is the value of the crop and the current market, what percentage has been destroyed or may be destroyed, and what is the cost of preventive measures to forestall potential destruction. Unfortunately, each of these measures can be assessed accurately only with the greatest of difficulty. Assigning a price value, for example, is a complicated business. Under free market conditions, the law of supply and demand dictates that an increase in production will be followed by a reduction in price. The effect of the reduction in price will differ, depending on whether a crop falls into the economists' category of "elastic" or "inelastic" demand. If the price of peaches falls because the crop was especially bountiful, the demand will increase. On the other hand, a drop in the price of potatoes (an inelastic demand) does not stimulate an increased demand, and oversupply actually depresses the market.

To be able to calculate profit and loss, we must have some measure of the inputs to agricultural production, the demand for products, and the supply of products. Inventorying these factors can be quite a guessing game. Economists usually divide the inputs into three categories: land, labor, and capital. At the present time, it is estimated that there are from 1.4 to 1.5 billion hectares (one hectare equals approximately 2.5 acres) under cultivation in the world today, a total of 1.8 billion hectares available, and 2.8 billion in permanent meadows and pastures. The value of land and its use depends on location, population density, farm size, forms of tenure, land prices, alternative use (in England, for example, 20,000 hectares of the best land is lost annually to urban development), and land management (when the basic properties and needs of land are not respected, there is massive deterioration, as exemplified by the great dust bowls created in the midwestern United States and the eroded wastelands that replaced the lush grasslands of East Africa when they were overgrazed). Labor is always difficult to evaluate, especially in agriculture, because employment is seasonal. Capital is also an elusive variable. It includes not only physical assets, land, buildings, equipment, stock, fertilizer, feed, unsold produce, etc., but also money owned or borrowed. Sources of financing, credit rating, all enter into this calculation.

Additionally, in trying to calculate benefits from control measures, we must take into consideration not only the *private* benefits to the farmer but also the social benefits as exemplified by the mythical county of Arooskegan. When imports from off-farm products replace land and labor, the extent of the benefits depends on the opportunities for other uses of the released resources. This is seldom taken into consideration in trying to evaluate profit and loss. Furthermore, benefits are seldom assessed over the entire lifetime of the program. For example, the development of resistance to insecticides is a form of depreciation. The entire problem has a time dimension.

Or consider the "real" damage suffered by a standing crop. What density of insects can inflict loss? How much destruction actually constitutes loss? How do we measure each of these variables? Chiarappa and his colleagues observed that lack of reliable data has created considerable embarrassment. They pointed out, as an example, that in 1963 the administrators of the University of California found that no such data were available and therefore requested a statewide study of crop losses. Also recognizing the need, the United States Department of Agriculture conducted a comprehensive study; however, the surveys are really quite unreliable because they depend heavily on the subjective estimates of individual observers, each with his own particular bias.

Steps are being taken to bring accuracy and objectivity to loss assessment by the establishment by the Food and Agriculture Organization of an international collaborative program for the development of reproducible methods for the assessment of crop losses. In the meantime, available figures can be used to prove almost any point, and the point they have most often been called upon to support is that insects are causing losses that cannot be tolerated. But the losses are economic, not nutritional. Keeping this firmly in mind, let us look at some of the estimates that have been published. These selected statistics are taken from the United States Department of Commerce Census reports (1965-70) and the United States Department of Agriculture Cooperative Economic Insect Reports for the years 1965, 1967, and 1970.

SOME STATISTICS

The third most important crop in the United States, exceeded only by corn, wheat, and soy beans, is the forage crop alfalfa. About 29 million acres are devoted to its cultivation. It constitutes fifty percent of the total hay crop. Among those insects feeding on it to a marked degree are: Egyptian alfalfa weevils, pea aphids, meadow spittlebugs, alfalfa caterpillars, beet armyworms, fall armyworms, potato leaf hoppers, and alfalfa snout beetles. The most formidable attacks in the past decade have been delivered by the alfalfa weevil. Because of it, planting has been reduced from 10 to 86 percent in various states to a national total of 3.3 million acres. The damage is set at $220 million.

In 1966, 2,822,600 acres of arable land were devoted to the culture of fruits and nuts, yielding 2,720,000 tons of apples; 149,000 tons of apricots; 149,000 tons of peaches; 616,000 tons of pears; and 399,000 tons of prunes to a value of $551,094,000. In the production of these crops, 15,806,000 pounds of insecticides and miticides costing $112,900,000 were used. In 1968, the value exceeded $700 million, the loss to damage, $78 million, and the cost of treatment, $29 million.

Citrus accounted for 2.9 percent of all food consumed in the United States in 1968 and 32.6 percent of all fruits consumed. The crop was valued at $664,682,000. In 1966, 10,384,000 pounds of insecticides and miticides were used to control its pests. In California, the produce of the 209,820 acres planted in citrus was valued at $178,375,000. In 1970, 19,790,000 pounds of toxicants, including 47 different kinds, were applied. Losses (reduction in yield and cost of chemicals) totalled $22,600,000. In Florida, where 950,000 acres are grown, $45-$60 million dollars were expended for pest control. The cost of insecticides for grain sorghum in parts of Texas increased from $100,000 in 1967 to $14,000,000 in 1970 with no increase in acreage. Insects are held accountable for growth loss and mortality of pines yielding saw timber to the extent of 20 percent, of which bark beetles account for 12 percent, giving annual losses of five billion board feet. Until 1965, $2,500,000 were spent on controlling the mountain, western, and southern pine beetle.

Losses to timber are obviously not a matter of starvation, and here the appeal is clearly in terms of profit and aesthetics. The extent of losses to timber and ornamental trees is particularly difficult to assess. When, as for example, gypsy moths defoliate trees, the visual result is there for everyone to see. The average person assumes that the tree is killed in every case, and there is pressure from all sides to spray. As a matter of fact, there is very little information on the effects of defoliation. In theory the most important effects are: mortality, loss of growth, and increased susceptibility to disease and attacks by insects that would normally attack sound trees. Collectively these effects are called "growth impact."

The deceptive aspect of assessing losses in terms of the collective term is that it embodies too diverse results which have different significance to different interests. The householder or municipality is much less concerned about temporary growth loss than he is about mortality. The lumber industry is interested in growth loss. Moreover, of all the things that insects can do to a tree, defoliation, the most psychologically disturbing to the citizen, is not the worst that can happen. In 1952, only fifteen percent of insect-related growth impact was due to defoliation. As one specialist, H. M. Kulman, has pointed out, economic analyses of potential damage to stands due to defoliation are scarce. In all the professional literature there exist only three small reviews on the subject. Yet in the face of this the lumber industry was reported in the *New York Times* of October 25, 1973, as demanding a lifting of the ban on DDT. Finally in 1974 private timber and chemical interests joined with the Forest Service in pressuring the Environmental Protection Agency into granting the Forest Service permission to spray 500,000 pounds of DDT on 650,000 acres of Northwest Forest lands to kill a tussock moth population that was already on the decline (*Sierra Club Bulletin*, April, 1974, p. 17).

The importance of defoliation is extremely hard to measure in any case because the age and location of trees, the stage of leaf development at the time of attack, and the degree and location of the loss of leaves all affect growth and mortality. Moreover, before blaming mortality on defoliation one should know that "normal" mortality is a difficult thing to assess. Finally, the effects of defoliation are different from one species of tree to the next. In general, evergreens and conifers usually survive a single complete defoliation in spring before the elongation of new foliage. Later, defoliation that includes both new and old foliage often, but not invariably, causes mortality. Most deciduous trees survive several defoliations and produce new sets of leaves in the same season. Conclusions from extensive studies of Connecticut forests during the period 1959-71 are that a single defoliation will not cause increased mortality over a ten-year period, and repeated defoliations cause an increase in mortality that is not more than doubling over a decade. Large-scale destruction of forests in Connecticut by defoliation seems unlikely (22).

But, you might say, the farmer or forester certainly knows how much he made on his crop in a good year and how much he made in a year that had heavy infestation, and the difference is his loss. In the years immediately following the Civil War, a comparison was made between the average number of bales of cotton produced in a fifteen-year period. This was compared with the average number of bales produced during years when there were heavy infestations of cotton worm. The percent of loss was calculated as 15.5 percent and (calculating $50 per bale) was recorded as approximately twelve million dollars. This dollar estimate, however, did not take into account fluctuations in price, reduction in transportation, ginning, and other costs, nor the demand at the time. The calculation of loss was not a matter of simple arithmetic in the 1860s nor is it now.

Chiarappa, Chiang, and Smith indicated that assessing actual crop damage is a grossly inaccurate pastime. Another agricultural economist, Headley, has indicated that data on resources allocated to agriculture are not very reliable either. In 1966, 353 million pounds of insecticides (not including sulphur and petroleum) were used for agricultural purposes in the United States alone. The value was $561 million. These figures represent one half of all pesticides used and equal 2 percent of total farm production expenses in that year. The figures do not include capital expenses. For example, in 1966, American farmers owned 1,016,000 power sprayers and spent $58,921,000 for custom pesticide application services, making a total annual expenditure (not including labor, etc.) of approximately $700 million dollars. Figures for another year, 1967, show that 1,615 teachers and research workers were employed in agricultural experiment stations plus 192 workers in extension services. At the same time, the budget for the fiscal year of the Entomology Research Division of the United States

Department of Agriculture was $16,680,230 of which only $1,008,900 was allocated for research on biological control.

It is apparent that a sizeable fraction of the alleged loss accrues from the cost of insecticides and their application. Since the farmer tends to adopt measures that he hopes will reduce the risk of damage without knowing whether that damage will actually occur or to what extent, a great deal of his control is in the form of "insurance." He tends to treat whether treatment is actually necessary or not, and to overtreat for good measure. The premiums can be very high. Thus, estimated losses do not represent actual damage by insects but instead are partly the cost of ignorance—the cost incurred by not being able to ascertain accurately how much damage an insect will inflict. They also represent self-inflicted losses that occur as a consequence of side-effects of massive bludgeoning of crops with wide-spectrum insecticides.

The fact is that damage to crops by insects is seldom complete, seldom of pandemic proportions, and seldom of yearly occurrence. For example, chinch bug damage to wheat in Illinois in 1888-89 varied all the way from 0 percent to 100 percent. Or consider rice. One fifth of all the world's arable land that is sown with cereals is sown with rice, and one half of the world's population lives on it. Its principal insect pests are bugs, stem borers, armyworms, and grasshoppers, all of which are of local and sporadic occurrence. According to one expert (D. M. Grist, 1965), there has been no serious threat by pests to the total crop in any land. On a world basis, borers, usually the most destructive insects, cause a 2 percent loss and bugs, a 5-25 percent loss of a local crop.

A "feel" for the great fluctuations in damage in terms of frequency, severity, and spatial extent can be gauged by thumbing through a very complete summary of pest insects of annual crop plants in Canada compiled by the Pestology Centre of Simon Fraser University. A few cases picked at random document the geographical and temporal variability: Periods of abundance of the pale western cutworm alternate with periods of relative scarcity; levels of damage by flea beetles are correlated with weather; the red-backed cutworm fluctuates and may be scarce in one province while abundant in another; the corn borer is a serious pest in the east and only a minor pest in the Prairie Provinces; the tobacco hornworm is a sporadic pest; in forty years there were eight serious outbreaks of the diamondback moth in prairie provinces; and so on. If there were more knowledge of the ecology and population dynamics of so-called pest insects, there would be no need for "insurance" spraying. How often do the premiums exceed the losses?

Modern agriculture has emerged as an industry. Agriculture and economics are tightly coupled, and the farmer, as the entomologist, Philip S. Corbet of the Canadian Department of Agriculture, has pointed out, is locked into a given system. At the present time, economics rather than pests seem to hold the key to world satiation. The complexity of providing enough food for the world has been succinctly stated by J. G. Harrar, President of the Rockefeller Foundation, who two decades ago led the foundation-supported groups that with Mexican scientists and their government effected a dramatic revolution in Mexican agriculture:

Appropriate soil and water management, improved plant varieties, and the use of fertilizers with other agricultural chemicals in conjunction with adequate systems of credit, transportation, and marketing, guarantee the possibility of vastly increased world food supplies. When buttressed by an enlightened leadership, growing investment, and a continually improving network of educational and research institutions, ultimate success is almost inevitable. The difficulty is to assemble the several critical components in appropriate terms of time, dimension, and place to bring about their necessary interaction.

There is, however, no unanimity as to the future course that agriculture should take. One economist (Capstick) sees it this way:

It is certain that, with modern techniques, farmers in the countries of western Europe and America can produce far more food than those countries as a whole can consume. For the world as a whole the picture may be different; *this we do not know, for while demographers are unanimous in predicting a steep rise in the world population, agriculturalists are far from unanimous in assessing the capacity of the world's agricultures to feed the new mouths even in the low standards which prevail over much of the presently underdeveloped world* [italics are mine]. On the one hand, therefore, it seems immoral to attempt to stem the flood of agricultural production in the Western world, since we cannot say that in the future the "surpluses" will not become, as they did in 1966 and 1967 [and 1975], very necessary stockpiles. On the other hand, the developing countries will not choose to be indefinitely receivers of food aid if they can, from their own resources, produce enough to provide an acceptable standard of nutrition for their peoples. . . .The possibility of the developing nations becoming customers in world markets for food produced in the West is also doubtful.

The problem which agricultural policies must attempt to solve is, therefore, not only that of withdrawing resources from agriculture but also that of replacing agriculture, or supplementing it, to provide a secure social and economic base for rural areas.

Whatever the future, whatever the country, the farmer is motivated to maintain a high sustained yield with the least energy expenditure. The insect is alleged to be a serious impediment to the realization of this goal. How serious an economic threat he is, is not known with any degree of certainty because reliable methods for estimating damage are not in use, losses are compounded by man's ignorance and ineptitude, and losses are extraordinarily subjective and biased. Thus, anyone can conjure up figures to support any crusade he wishes to espouse. Most often the statistics have been wielded in the service of the pesticide industry with the threat of famine held over the heads of those whose allegiance wavers. The insect does indeed interfere in our relations with plants. If we wish to maintain the ecologically and evolutionary artificial relation that we have with domesticated plants, we must certainly do something about some insects, but we should not be misled as to our motives or deceived with respect to the aggressiveness with which we must fight.

The insect does not compete with our bellies; he competes with our pocketbooks. What we do should be based upon biological wisdom and not motivated solely by the philosophy of business.

References

1. Abelson, P. H., The World's Disparate Food Supplies. Science, 187 (4173).
2. Aldrich, D. G. (ed.), Research for the World Food Crisis. Publication No. 92, Amer. Assoc. Adv. Sci., Washington, D.C., 1970, 323 pp.
3. Anon., The World Wheat Situation and the International Wheat Agreement. The International Wheat Council, London, 1954, 34 pp.
4. Anon., The World Food Situation and Prospects to 1985. Foreign Agricultural Economic Research Report No. 98, Economic Research Service, U.S. Dept. Agri., Washington, D.C., 1974.
5. Anon., Review of World Wheat Situation 1968. International Wheat Council, London, 1968, 79 pp.

6. Beirne, B. P., Pest insects of annual crop plants in Canada I-III. Mem. Ent. Soc. Canad., 78 (1971), 1-124.

7. Beirne, B. P., Pest insects of annual crop plants in Canada IV-VI. Mem. Ent. Soc. Canad., 85 (1972), 1-73.

8. Bell, D. E., Responsibilities of government in the support of food, production and distribution. In: *Prospects of the World Food Supply*, National Academy of Sciences, Washington, D.C., 1966, pp. 3-10.

9. Capstick, M., *The Economics of Agriculture*. Geo. Allen and Unwin, London, 1970, 163 pp.

10. Chiarappa,. L., Chiang, H. C., and Smith, R., Plant pests and diseases: assessment of crop losses. Science, 176 (1972). 769-73.

11. Corbet, P. S., Application, feasibility and prospects of integrated control. Symposium 14th International Congress Ent. Canberra, Australia, 1972.

12. Entomological Society of Canada, Pesticides and the Environment. Brief prepared at the request of The Board of Governors—1970. Suppl. Bull. Ent. Soc. Canad., 3 (1970), 1-16.

13. Grist, D. H., *Rice* (4th ed.). Longmans, Green & Co., London, 1965, 548 pp.

14. Gunther, F. A., and Jeppson, L. R., *Modern Insecticides and World Food Production*. Chapman and Hall, London, 1960, 284 pp.

15. Harrar, J. G., Chairman's introductory remarks. In: *Prospects of the World Food Supply*. National Academy of Sciences, Washington, D.C., 1966, p. 2.

16. Headley, J. C., Economics of agricultural pest control. Ann. Rev. Ent., 17 (1972), 273-86.

17. Knapp, J. B., The role of international agencies in aiding in world food production. In: *Prospects of the World Food Supply*. National Academy of Sciences, Washington, D.C., 1966, pp. 11-17.

18. Kulman, H. M., Effects of insect defoliation on growth and mortality of trees. Ann. Rev. Ent., 16 (1971), 289-324.

19. Malenbaum, W., *The World Wheat Economy 1885-1939*. Harvard University Press, Cambridge, 1953, 262 pp.

20. Rabb, R. L., and Guthrie, F. E., *Concepts of Pest Management* North Carolina University Press, Raleigh, 1970, 242 pp.

21. Seaborg, G. T., Science, technology, and development: A new world outlook. Science, 181 (1973), 13-19.

22. Stephens, G. R., The relation of insect defoliation to mortality in Connecticut forests. Bull. Conn. Ag. Exp. Sta., 723 (1971), 1-16.
23. United States Department of Agriculture, Third Report: the United States Entomological Commission, Washington, D.C., 1883, 347 pp.
24. United States Department of Commerce Census Reports, 1965-1970, Washington, D.C.
25. United States Department of Agricultural Cooperative Economic Insect Reports (1965, 1967, 1970), Washington, D.C.
26. Wade, N., World food situation: pessimism comes back into vogue. Science, 181 (1973), 634-38.
27. Wressell, H. B., A survey of insects infesting vegetable crops in southwestern Ontario, 1969. Proc. Ent. Soc. Ontario, 101 (1970), 13-23.

Chapter Four
Man's Own Plague

4
Man's Own Plague

For centuries, man has relied upon faith to curb the appetites of plant-feeding insects—originally faith in the efficacy of prayer and legal sanctions and later faith in the miracles of DDT and its successors. In Biblical times, appeals were directed to Jehovah; in Roman times, special offerings were made to propitiate the gods responsible for pests. In medieval times, legal action and excommunication were the means of redress. Prayer was intended as preventive control, whereas jurisprudence was *ex post facto* control. Legal action occurred frequently, with the result that a complex codification of procedures eventually evolved. A Burgundian, Bartholomaeus Chassanaeus, wrote a treatise specifying the rules under which a suit could be entered in court against grasshoppers (28). A court would be convened upon written request, a judge appointed, and a prosecuting and a defense lawyer assigned. The prosecutor would present the case against the grasshoppers and demand that they be found guilty and burned. The defense would argue that the demand was illegal because first the grasshoppers had to be requested to leave the country within a specified period of time. If at the expiration of this period they had not left, the proper sentence was excommunication.

As might be expected, not everyone in the legal profession agreed on the letter of the law. Years later, Chassanaeus' brief was challenged by another jurist, one Hiob Ludolph, who pointed out that before any action could be taken, the accused grasshoppers had the legal right to be summoned four times! If they ignored these summonses, then they could be forcibly brought to court whereupon the suit could proceed. There is a glimmering of ecological relationships here because it is further stated that other interested parties, to wit the birds which feed on grasshoppers, have a right to be heard. Furthermore, great injustice would be done by extraditing the grasshoppers to innocent adjacent territories. Last but not least, there was no provision in canon law for ecclesiastical action against grasshoppers.

Justice was not to be obstructed, however, by mere legalistic semantics. A typical suit was that brought to court against worms in 1479, in Berne, Switzerland. Despite an able defense, the cutworms were pronounced guilty, duly excommunicated by the archbishop, and banished. A contemporary chronicler added an interesting postscript to the case: "No effect whatsoever resulted, evidently on account of the great depravity of the people."

As insurance, the people took further action by attempting to scare grasshoppers away. They marched upon them waving their arms or sticks or whatever they could lay their hands on, meanwhile creating a din by shouting or beating drums. This physical approach lasted through colonial times in America and persists in rural areas of Africa and China. Generally speaking, the most direct method of ridding crops of insects was by scaring them away or picking them off by hand. As late as 1862, these methods remained the choice ones against many insects. Excerpts from Harris' *Treatise on Insects Injurious to Vegetation* makes fascinating reading: June beetles, shake from the tree in the daytime; rose beetles, hand picking; wireworms, collect from sliced potatoes; beetle larvae boring in trees, protect the woodpecker; turnip flea-beetle, sweeping; blister beetles, shaking into pans; grasshoppers, drag sheets; squash bugs, early hand picking; pear psylla, hand picking; parsley swallowtail caterpillar, hand picking; hairy caterpillars, pay children to collect them by the quart; tent caterpillar and white tussock moth, hand destruction of eggs in the winter; and so on.

One of the most ingenious methods of control was that devised by a Colonel T. Forest of Germantown, Pennsylvania, where curculios (boring larvae) were injuring plum and apricot trees: "Having a fine plum tree near his pump [he] tied a rope from the tree to his pump handle, so that the tree was gently agitated every time there was occasion to pump water. The consequence was, that the fruit on this tree was preserved in the greatest perfection."

Prior to the nineteenth century, insect control was not on an organized basis. Despite the occasional plagues of locusts and cutworms, the greatest deterrents to successful harvests were, as we have seen, bad weather, lack of fertilizer, and restrictive socioeconomic conditions. As the Agricultural Revolution progressed, these problems receded into the background and damage by insects attracted more attention. Some control in the form of crop rotation and timing of planting may have existed. Probably the first written advice in the West on how to deal with farm insects was that given by Linnaeus (26) in 1735. Among the treatments that he suggested were: linen cloth soaked in whale oil and wrapped around tree trunks to deter caterpillars, fumigation with fumes of rancid oil to kill codling moths, the introduction of lady beetles and parasitic wasps into orchards to control aphids, and the introduction of predatory beetles to kill caterpillars. In 1795, various articles by Peck (18) appeared in farm journals. In 1837, Kollar (18) published a book on farm insects. In general, most of the remedies of the era were discovered by European and American gardeners and published for the benefit of the public in local newspapers, pamphlets, and various agricultural journals.

In 1841, the entomologist Thaddeus William Harris, probably the first American entomologist to receive public compensation for his labors, later librarian of Harvard College, summarized most of the existing knowledge on both sides of the Atlantic although he himself was only familiar with agriculture in the vicinity of Boston and had traveled little. From his report, now a classic, we learn of some of the earliest uses of chemicals, or, more properly, decoctions, concoctions, elixirs, and exotic formulae that began to supplement hand picking or were used in conjunction with it. As there was no point in removing insects from plants unless they were prevented from returning, one picked, and then killed. Picking was laborious and time-consuming. Much effort would be saved by removing and killing simultaneously or by killing on the plant and not worrying about removing corpses that did not fall off. Better still, prevent the insects from coming to the plants in the first place. Solutions painted, sprinkled, or watered onto plants could cover foliage efficiently. If plain water could remove some insects, how much more effective water mixed with something noxious. Here an anthropomorphic attitude helped; materials that were noxious or toxic to people were added to the water. Remedies were compounded from materials commonly available.

The Town Records of Northfield on Staten Island, New York, record for the period 1783-1823 that "soft cow dung put in water and elder sprouts bruised and steeped in water put over any plant prevents any insect injuring them (27)."

During the same period it was recommended that cucumbers

be protected from cucumber flies with an aqueous mixture of tobacco and capsicum (red pepper) sprinkled over the vines (26).

Other curious remedies (28) included infusions of walnut leaves against flea beetles, strong soapsuds, potash water, whitewash plus glue, whale oil soap plus Peruvian guano, and solutions of whale oil soap for various species of leaf-feeding insects. Dusts were also employed, being sieved on plants from sieves attached to the ends of long poles. Among those used were: ground plaster of Paris, charcoal dust, powdered soot, sulphur plus Scotch snuff, and air-slaked lime. For borers, camphor in plugged holes was recommended. The wooly root-louse of apple was attacked with a mixture of melted resin and fish-oil painted on the trunks and large roots. The recommended treatment for cutworms was soaking grain before planting in copperas water (iron sulfate), rolling seed in lime or ashes, mixing salt with the manure, fall plowing of sward lands intended for wheat or corn the following year, collecting larvae by hand, manuring soil with sea mud, protecting cabbage plants by wrapping a walnut or hickory leaf around the stem. Cankerworms were combatted by banding trees with clay mortar, strips of old canvas, or strong paper tarred, troughs of tin filled with cheap fish-oil, melted Indian rubber, jarring the trees, or using pigs to destroy pupae under the ground.

There is every indication that, grain excepted, the acreages under cultivation in America during the eighteenth and early part of the nineteenth centuries were individually modest. With some exceptions, the amount of damage inflicted by insects was not known; but, since insects had obviously destroyed some gardens and some orchard trees, it was clear to the agriculturalists of the times that insects could be dangerous, or, if not dangerous, at least a nuisance. It was probably about this time that the concept of insects as pests had its genesis. In America, at least, the public consciousness of the insect as a pest, if it had not already come to the attention of the practical farmer first-hand, was stimulated by the increasing number of appearances in print of accounts of insects on crops by men like Harris, who were first and foremost interested in the insect itself. Insects were noticed more and more for their own sake. The fact that they also lived on, and sometimes damaged, crops only made them the more interesting.

It is probably fair to say that during this period the accelerated search for remedies was more closely correlated with increasing interest and knowledge of insects than with increasing damage to crops. One book published in America in 1891, written by C. M. Weed, listed practically every known plant-feeding insect as a pest. In any case, the various prescriptions in use at the time were only marginally effective and were unpleasant to apply. Whenever new materials came into general use for other purposes, they were tried on insects. Thus, when kerosene replaced whale oil and fish oil as fuel for lamps, it was tried against insects.

This was a lucrative period for charlatans and quacks. Just as patent medicines and kitchen-concocted elixirs were the stock in trade of every apothecary, drummer, and peddler at local fairs, so were remedies for use against insect pests real or imagined. If an insect was not a pest before an itinerant peddler began his pitch, it certainly was afterward. Entomologists rebelled.

In his reminiscences of the history of applied entomology, L. O. Howard quotes the following excerpt from the editorial in the first number of "The Practical Entomologist," 1865:

The agricultural journals have from year to year, presented through their columns, various recipes, as preventive of the attacks, or destructive to the life of the "curculio," the "apple moth," the "squash-bug," etc. The proposed decoctions and washes we are well satisfied, in the majority of instances, are as useless in application as they are ridiculous in composition, and if the work of destroying insects is to be accomplished satisfactorily, we feel confident that it will have to be the result of no chemical preparations, but of simple means, directed by a knowledge of the history and habits of the depredators.

The last sentence is particularly significant because it stated a wisdom that until relatively recent times was largely forgotten.

With the exception of sulphur and later arsenic (first used by the Chinese in the sixteenth century), the first insecticides worthy of the name were of botanical origin. Without realizing their chemistry, man began to utilize in crude form the compounds, mostly alkaloids, that some plants had evolved as defenses against herbivores. Infusions of tobacco were first used in 1690 in France for lace bugs on pear trees. Tobacco was in common use as an insecticide in the eighteenth century. Plants of the genera *Derris*, *Lonchocarpus*, *Millettia*, *Mundulea*, and *Tephrosia* were used from time immemorial as fish poisons. Eventually, rotenone and related alkaloids were extracted and characterized and introduced into modern use in the 1920s. White hellebore was used in France as early as 1787 against aphids. It is derived from the lilies *Veratrum alba* in Europe and *V. viride* in the United States. The greatest of them all, pyrethrum, probably has an ancient history. A daisy-like plant of the family Compositae, it grew wild in the Caucasus mountains, where it was known variously as flea-grass, flea-killer, and Persian camomile. Sometime around 1807, an Armenian merchant, Jumtikoff, noted the effectiveness of a prepared powder and sent the information to his son, who accumulated a quantity of powder for export in 1828. It was introduced as an insecticide into the United States by C. V. Riley in 1885.

One of the earliest strictly synthetic insecticides came into use about this time. In France, mildews, phylloxera, and other epidemic diseases were striking at the vineyards. Two remedies were discovered, Bordeaux mixture (hydrated lime plus copper sulfate) and Paris Green (copper acetoarsenite). Who first used Paris Green in America is not known, but it came into use against the Colorado Potato Beetle sometime in the 1860s, when the beetle was dispersing eastward at an alarming rate. The standard treatment consisted of dusting with hellebore and ashes and then removing the larvae by shaking the plants. On May 28, 1869, a George Liddle wrote to the editor of the Galena (Illinois) *Gazette* that since 1868 he had been using Paris Green successfully against the beetle. He mixed one pound of Paris Green with two pounds of flour, and sifted it through coarse muslin cloth onto the plants while they were still covered with dew. Three pounds spread over an acre caused larvae to fall to the ground by the thousands. By 1872, its use had spread to other insects.

In 1880, a Michigan farmer by the name of Cook discovered that kerosene mixed with a solution of soap made a stable emulsion which did not kill foliage as did the raw oil. By the 1860s, many chemical formulations, usually prepared by the farmer himself, were in use. The active ingredients were compounds of antimony, arsenic, mercury, selenium, sulphur, thallium, or zinc. Hydrocyanic gas, first discovered in 1782, found its first use as a fumigant in 1880 in citrus groves. The stock list of insecticides was now complete for many decades to come, and, by 1934, America was using annually 1,000,000 pounds of rotenone, 30,000,000 pounds of sulphur, 7,224,000 pounds of arsenicals, 10,000,000 pounds of pyrethrum, and 4,000,000 pounds of Paris Green.

Insecticides were here to stay. Insects could be killed most
efficiently by chemicals. More important still, one could see the
creatures die. No doubts remained. There was money to be made
by manufacturing them, and the greater the volume that could be
sold the better. There is no indication that the supply was tailored
to the demand, but, rather, that the demand was carefully nurtured.
Two new factors entered into man's relations with insects at this
point: the growth of commercial producers of insecticides and the
appearance of the professional economic entomologist. It is an
interesting aside that of all the branches of zoology, entomology
is the only branch concerned with the destruction of the animal
that it studies. The position of entomology vis-à-vis other zoological sciences was described in an address to the Entomological
Society of America in 1967, when it was remarked that entomology had advanced a long way from "the science that treats of
insects" (as described in the eleventh edition of the *Encyclopedia
Britannica*) to the "profession that controls insects" (as described
in a Plenary Symposium in 1966 under the title "Entomology
looks at its Mission"). The words attributed to Archie the cockroach by Don Marquis describe the essence of contemporary
entomology:

> i thought of all
> the massacres and slaughter
> of persecuted insects
> at the hands of cruel humans
> and i cried
> aloud to heaven
> and i knelt on all six legs
> and vowed a vow
> of vengeance.

In all justice, it must be pointed out that of all groups of animals
insects as a whole are the most destructive to our crops, homes,
livestock, belongings, and persons.

THE COMING OF DDT

The crisis that was to develop, the plague that we were to loose
upon ourselves, was born in World War II. The medical departments of the Army and Navy realized early in 1941 that American
troops would be exposed as never before to a formidable array

of arthropod-borne diseases. Whereas the chief miscreants in World War I were lice, fleas, and bedbugs, there was now a galaxy of insects that transmitted tropical diseases: mosquitoes, biting flies, lice, and ticks. Among these diseases were several forms of malaria, sleeping sickness, relapsing fever, sandfly fever, dengue, typhus, encephalitis, and filariasis. Available to combat the vectors were methyl bromide for fumigation against lice, pyrethrum as a powder for personal use, and oil to spray on bodies of water where mosquitoes were breeding. The means to combat these insects were inadequate; but the exigencies of war demanded a quick and massive kill.

Upon recommendation of a war-time committee of the Office of Scientific Research and Development, money for research and development was made available to the Bureau of Entomology and Plant Quarantine of the United States Department of Agriculture and the Gorgas Memorial Institute of Panama. The former conducted most of its field work and screening tests in Orlando, Florida. To this laboratory searching for effective insecticides, manufacturers sent hundreds of chemicals for routine testing. One day the New York branch of the Geigy Company of Basel, Switzerland sent a small package labeled "Gesarol." It was tested routinely along with hundreds of others. The testers quickly discovered that this material had remarkable properties. Minute amounts were lethal, not only to lice, the principal objects of the investigation, but also to mosquitoes, bed bugs, fleas, flies, and indeed to nearly every insect. At one point, a small amount of the chemical accidently blown into the air almost destroyed the entire colony of yellow-fever mosquitoes in the laboratory. Analysis of Gesarol by chemists of the Department of Agriculture revealed that the active ingredient was dichloro-diphenyl-trichloroethane —DDT. The panacea had been found.

The success of DDT spurred organic chemists to greater efforts, and soon other lethal chlorinated organic compounds were marketed. Many became household words: lindane, dieldrin, methoxychlor, chlordane, heptachlor, etc. During roughly the same period, research begun in the 1930s in Germany culminated in the synthesis of a new category of extremely toxic chemicals, the organophosphates. Much of the research was motivated by a desire to develop chemical warfare agents, and some of the first compounds were lethal nerve gases. During the war, great secrecy surrounded the work on these compounds. Two of the earliest with potential as insecticides, HETP and TEPP, were patented in Germany in 1942, and in the United States in 1943. Two were added to the household vocabulary: parathion and malathion.

The spectacular success of DDT under war-time conditions and its later conversion to peacetime agricultural uses stimulated the insecticide industry to gargantuan efforts. Within a remarkably short time, DDT was joined by even more potent synthetic insecticides, the organophosphates just mentioned. The objective now became the total and permanent destruction of insects. The word "eradication" replaced the word "control."

The idea that an insect could be totally eliminated first found expression in the later nineteenth century, at which time entomologists were divided on the soundness of the idea. Some thought it feasible; others impossible. As late as 1958, some eminent entomologists were still convinced of the feasibility of eradication; furthermore, there developed an unshakable faith that extinction on a national or state scale was possible. "It is better to live without than learn to live with destructive insects." But a more sober assessment shows that there is no hope of making the insect extinct (although we have succeeded well with the dodo, the passenger pigeon, and larger beasts).

When the West African mosquito *Anopheles gambiae*, the principal vector of malaria, stowed away on aircraft returning to South America and established itself in Brazil, a massive campaign to eradicate it began. It was said to have been completely eradicated. Mediterranean Fruit Fly arrived in Florida in 1929, and was said to have been eradicated three months later—the first of several times. Unfortunately, insects are unaware of political boundaries. Application of insecticides took place over thousands of acres, frequently by spraying from aircraft. The use of insecticides became automatic, and was not cued to damage actual or potential. Sprays were applied by the calendar and as insurance. It is so simple to douse an area and not have to worry about insects. Growers were urged on by the highly competitive insecticide industry, and often by extension entomologists (those working at experiment stations or as field agents), many of whom had little appreciation of ecology or of the insect as an organism.

A farmer struck by an insect pest more often than not will get advice from the insecticide salesman. Salesmen outnumber specialists and extension entomologists. They are supported by efficient organizations that want to increase sales. Where else is the farmer to turn? As C. B. Huffaker of the University of California (California uses 10 percent of all the insecticides sold in the United States) remarked, "pesticide salesmen have become the major source of grower advice. Experiment Station and Extension Service personnel do not have the personal day-to-day contact with growers. Pesticide usage is promoted sometimes under great pressure to sell, irrespective of need or possible later consequences, and often in ignorance at least of the ecology of the ecosystem that is often disrupted."

The number of amateurs who know only non-selective pesticides far outnumber the specialists who know which pesticide to use and are developing and applying improved and selective control methods. Speaking of DDT alone, it is estimated that roughly 4.5 million tons have been manufactured since its discovery. In 1958, just to select a year at random, 142 million pounds of insecticide were produced. In 1963, the figure was 169 million pounds. More than 80 percent of pesticides manufactured in the United States are synthetic organic compounds valued at 750 million dollars a year. Merchandising has replaced biology.

The public has been brainwashed into believing that man and insects can not co-exist. All insects are cast in the role of pests even though only one-tenth of one percent of all species compete with man, and the prevailing philosophy maintains that all insects should be killed and that if one pound of insecticide was good two are better. An attitude unusual in industry was expressed by one pesticide company, Zoecon of California: "Voracious as locusts and other insects can be, the competition between men and insects is not nearly so great as might be expected. In general, the two groups coexist with relatively little friction, each dominant within its own considerable sphere of influence." Although, as we shall see later, the philosophy of control has changed from one of eradication to one of maintaining pests at a tolerably low level, the years immediately following the spectacular success of DDT nurtured a concept of "pest" that permeates all levels of contemporary thinking and is extremely resistant to eradication.

"Pest" is a subjective term. In the trade, it embodies birds, insects, mites, molluscs, nematodes, rodents, plant pathogens, and weeds alike. To many people, insects are synonymous with pests. This phobia of insects is by now a cultural attitude, and attitude and circumstance have a great deal to do with what constitutes a pest. A single hungry female mosquito on the prowl in a bedroom at night qualifies as a "pest," but does the lone caterpillar that consumes a few leaves from a favorite bush?

The average person is unaware of the presence of insects until they reach some critical number, at which point they are noticed and become pests. For example, aphids often become so numerous on certain trees, such as maples, that their excretions may fall on cars parked nearby. This will elicit a prompt communal response: The trees are forthwith sprayed with insecticide. The motivation, however, is not to save the trees from a parasite, actual or potential, but to keep the automobiles from being splattered with excretions. It is easier, though not necessarily less expensive, to petition the town to spray the trees than it is to wash the cars. What effect the spray might have on the trees or upon other forms of life is usually not considered.

This attitude toward insects was strikingly demonstrated in the Borough of Princeton, New Jersey, in the summer of 1970. In May and June of that summer, the seventeen-year cicada made its scheduled appearance. After having spent sixteen years in the ground, out of sight and out of mind, the nymphs began to emerge in their usual overwhelming numbers. They ascended the trunks of trees, bushes, signposts, sides of buildings—anything vertical —split their skins along the back, struggled free of the old vestments, changed from infantile white to their adult colors, and climbed or flew out of reach to the crowns of the shade trees. There they fed and courted and sang. And how they sang! Before long, the human population of Princeton was spraying furiously. And why? Feeding by cicadas does no damage worthy of consideration. The damage they do inflict, and it is negligible except to young trees in a nursery, is restricted to the terminal twigs, where the females slit the bark with their ovipositors to lay their rows of eggs. After six or seven weeks, the eggs hatch, the nymphs descend to the ground, where they disappear into the soil for another seventeen years. But the citizens of Princeton (and in this they are not unique) do not like to have swarms of cicadas around even for a few weeks. As one colleague remarked to me, "I don't mind them, but my wife cannot stand having them around the garden." While another whose house is located on the main route along which tractor-trailers blast most of the day complained that he could not stand the "infernal racket of the cicadas." So the trees were sprayed.

Not to belabor the point, I cite but one more example which is representative of a widespread attitude among many farmers who, more than anyone else, have a legitimate right to be wary of insects. In parts of Texas, peanuts have, in recent years, become a money crop. Thousands of acres of land too impoverished to support other crops have been sown with peanuts. As with other plants, several kinds of insects feed on the peanut foliage. Many do no damage, in the sense that their presence on the plants and consumption of foliage do not decrease productivity as measured in dollars. Yet the peanut farmer sprays, educated to a fine degree by the salesmanship of the insecticide representative.

The awakening has been slow in coming, and in retrospect, if people had used a little biological horsesense and had not been swept away by fantasies of insect-free living, chemical miracles, gain without effort, and handsome profit, the trouble might never have arisen. It cannot be claimed, as many apologists now maintain, that relevant biological knowledge was lacking. Stated flatly: It was not. Thirty years ago, ecologists knew all about food chains, a great deal was known about prey-predator relationships (even the medieval jurist was concerned about the rights of birds as interested parties in the man-insect relationship), and the persistence of DDT was not only recognized but extolled (said one writer: "Of the many insecticidal properties of DDT, probably the most exciting and important. . .is its residual action"). The phenomenon of resistance had been known since 1908, when the San Jose scale developed resistance to lime-sulfur applied to apple trees; the California red scale, an insect of citrus, developed resistance to fumigation by hydrogen cyanide in 1910. In 1914, the entomologist Melander asked: "Can insects become resistant to sprays?" As early as 1945, another entomologist, E. H. Strickland, in an article entitled "Could the widespread use of DDT be a disaster?" recognized and exposed the dangers. The modern farmer is in the same predicament that the medieval farmer was even though the reasons are different. Without realizing it, we embarked upon a catastrophic course.

RESISTANCE TO INSECTICIDES

The first indication that the age of miracles might indeed have passed came from a report in 1946 that houseflies in Sweden were no longer killed by DDT. The failure was blamed on an inferior grade of insecticide or its improper use. That explanation was short-lived. Reports began to accumulate from all corners of the globe. Flies, mosquitoes, lice, agricultural insects—all kinds— were not succumbing. If one reads the entomological literature of the times, he detects an underlying feeling of disbelief and astonishment. One had only to go back to the older literature, however, to learn that nothing new and bizarre was happening. The same thing had happened with old-fashioned insecticides: hydrogen cyanide, lead arsenate, lime sulfur, tartar emetic. What indeed was happening?

Geneticists knew exactly. They cited it as an example of rapid evolution. Resistance is a population phenomenon, not an individual phenomenon. Individuals are not becoming immune to an insecticide through constant or repeated exposure; instead, the ratio of susceptible to resistant individuals in a population is changing. In any given population, some individuals have genes that render them immune to an insecticide. These genes have always been present in the propulation, but, having no obvious selective value, they have been present in a few individuals only. Then along comes an insecticide that kills all of the individuals lacking genes that confer immunity. The only individuals left to breed are those with some degree of immunity. They generate a population with fewer susceptible individuals. In this way, as the application of an insecticide is continued, the population as a whole becomes less susceptible with each succeeding generation. The insecticide has acted as a selector of the fittest individuals, in the best Darwinian tradition. Increasing the dosage only serves to make the selection more rigorous, and an even more tolerant population. The population now has a high frequency of genes for resistance, and these are also genes that confer good viability and fertility under those conditions.

When treatment with insecticides is stopped for a few years, the return to susceptibility is slow, because there are few susceptible genes, and they have little selective advantage. When later the use of insecticides is resumed, resistance appears very rapidly, because a high frequency of genes for resistance is still present. Sometimes one insecticide causes the development of resistance only to itself, sometimes to many insecticides; sometimes a population resistant to one insecticide may then be made resistant to another without losing resistance to the first.

If we had been privy to the inner council discussions hastily convened thirty years ago by the National Research Council, the Office of the Surgeon General, the United States Department of Agriculture, and others, we would have learned that their astonishment was rapidly replaced by concern and a realization that something had to be done. Even then, however, the prevailing sentiment seems to have been that the solution lay in developing other insecticides to which insects did not become resistant. There was a feeling that the trouble lay not so much with the insect, that is, with the biological system, as with the chemical one. And indeed it is true that resistance develops more readily with some chemicals than with others. Time, effort, and money now went toward the compilation of records of resistance, evaluation of resistance, elucidating its generic basis, and searching for the biochemical basis. There was a hope in many quarters that a deeper understanding of the mode of action of insecticides together with an understanding of the biochemistry of resistance would point the way to the development of foolproof insecticides.

With all the evidence in hand, with faith in the efficacy of insecticides still intact, with immediate practical problems of the control of medically important insects pressing for attention, and with the spectacular success of DDT in controlling malaria in former theaters of war (such as the explosive typhus epidemic in Naples in December 1943 and January 1944 still fresh in mind) few people had any intuitive grasp of the basic biological significance of resistance and fewer still were prepared to look with any disfavor upon insecticides. Industrial, governmental, and university laboratories carried on research programs relating to all aspects of insecticide control.

At the time, there was little financing and sympathy for other forms of control. In universities, whole graduate departments and research programs were supported by monies allocated to control by insecticides. It was a simple matter in those days to find support for one's research, however basic and esoteric in nature, simply by referring somewhere in the application to chemical control. Long before there was a National Science Foundation, the various departments of the Armed Forces were supporting all the work that was being conducted in insect physiology because it was hoped that it would provide a chemical panacea. This was a tremendous boon to the study of basic insect physiology.

As time passed, however, it became increasingly apparent that it was in the biological nature of insects to develop resistance to insecticides. By 1967, resistance had developed in 224 species, of which 127 were crop, forest, or stored products pests. Substituting one compound for another and alternating one compound with another might solve the problem temporarily; but these procedures did not make the problem go away. Nine different types of resistance occurred. Heavier and heavier dosages of DDT and the new and powerful organophosphates were rained upon plants with no appreciable effect. Nothing can bring the problem home more forcefully than to visit a laboratory where cultures of resistant varieties are kept for study and to observe a thriving colony of flies living in a cage coated solidly with DDT. (In the meantime, the search for new compounds continues despite the certainty that they too will cause resistance.)

Soon the price of insecticides exceeded the profits to the farmer. The remedy began to cost more than the disease. As the figures quoted in the previous chapter illustrate, the farmer was faced with skyrocketing costs with no corresponding increase in profit. Part of the rise in cost was attributable to the increase in the cost of developing and producing insecticides; part to the heavier dosages required to combat resistance. Despite increased application of insecticides, there has been no reduction in crop losses; in fact, they have increased. Average total losses in 1904 were estimated to be about 11 percent. The figure for 1968 was 13 percent.

INDISCRIMINATE KILLING

A second crisis followed quickly upon the heels of the first. When insecticides killed, they killed not only the insects against which they were directed, but also decimated populations of other species as well as making way for new pests. Insects and other arthropods that had never before bothered growers became destructive overnight. As one entomologist remarked in 1952, "never before have so many pests with such a wide range of habits and characteristics increased to injurious levels following application of any one material as has occurred following use of DDT in applied spray programs." The so-called broad spectrum insecticides, and especially those that remain around for a long period, kill so many beneficial insects that the parasites and predators that normally keep other insects from becoming pests are removed from the ecosystem. This wholesale slaughter is illustrated by a comparison of the number of beneficial insects found in trees sprayed three times with DDT and those treated with the insecticide Rhyania—an insecticide of botanical origin. The trees treated with DDT had a residual population of 69 individuals representing six species of predators; the other trees, 817 individuals representing the same species.

A most striking picture of what may happen when resistance and destruction of beneficial insects are combined has been given by the entomologists Professors Richard L. Doutt and Ray Smith of the University of California (8), writing about cotton growing in the Canete Valley of Peru. The following example is a particularly good one because Canete Valley is isolated from forty similar valleys by the sea on one side and desert on the others: Cotton became the principal crop shortly after 1920, and from 1943 to 1948, cotton was treated with arsenicals and nicotine to control three pests of which the tobacco budworm (*Heliothis virescens* F.) was the most destructive. The others were the leaf worm (*Anomes texana* Riley) and the weevil (*Anthonomus vestitus* Bohm). In 1949, the budworm population exploded, an outbreak brought on by the introduction of insecticides that killed natural parasites, by the planting of ratoon cottons (second and third year cotton), and the elimination of all other vegetation that normally provided an over-wintering habitat for parasites and predators. The yield of cotton dropped from 415-526 lbs/acre to 326. A new and heavier spray program was adopted. DDT, BHC, and Toxaphene, the new organic insecticides, saturated the valley. Trees were even felled to facilitate aerial spraying. Within four years, yield nearly doubled to 648 lbs/acre.

Year by year the number of treatments was increased and begun earlier. By 1952, BHC no longer stopped aphids. By 1954, Toxaphene was no longer effective against the leafworm. In 1955-56, the weevil multiplied alarmingly. Then a new pest, the Peruvian leaf-roller (*Argyrotaenia sphaleropa* Meyrick) arrived on the scene. The budworm became resistant to DDT. The chlorinated hydrocarbons were replaced by the new organophosphates. From an initial program of treating every 8 to 15 days, spraying was stepped up to every three days. Five new pests attacked the cotton: another leaf-roller (*Platynota* sp.), a caterpillar (*Pseudoplusia rogationis* Guen.), and *Pococera atramenalis* Led., *Planococcus citri* Risso, and

Bucculatrix thurberiella Busck. These insects did not become pests in neighboring valleys where organophosphates were not used, even though they were present in small numbers.

By 1956, yields dropped to 296 lbs/acre and there was economic disaster. In desperation the growers begged for help. Their experiment station stopped the use of organic insecticides, forbade the planting of ratoon cotton and the cultivation of marginal land; parasites and predators were reintroduced; a cotton-free fallow period was required; only arsenicals and nicotene were allowed as spray. Within a year the "new" pests reverted to non-pest status and the yield rose to 468 pounds per acre. Between then and 1969, it fluctuated between 644 and 922 pounds per acre, the highest in the history of the valley.

This is not an isolated example nor are the problems restricted to the highly industrialized nations. They are equally vicious in developing countries. In Africa, cocoa was sprayed to cure minor problems; major problems developed. In Ceylon, tea was sprayed with dieldrin, and yields increased; then chaos struck. An internationally renowned authority on insecticides, Robert Metcalf, summed up the situation succinctly as follows:

For many years it has been recognized that the greatest single factor in preventing insects from overwhelming the rest of the world is the internecine warfare which they carry out among themselves. The entomophagous insects are generally inconspicuous and their absence is scarcely noticed until elimination by the ravages of climate or a blanket of insecticide allows the host species to increase suddenly toward its maximum biotic potential. Examples of this sort have been increasingly frequent as a result of the haphazard use of highly toxic and presistent organic insecticides, and familiar cases include the outbreaks of red spider mites . . .following applications of DDT, Dieldrin, Sevin, etc., and cottony-cushion scale, *Icerya purchasi* Mask., after drift of DDT and malathion.

The insecticide may kill the beneficial insect directly or indirectly by poisoning its food. Sometimes dosages that are too low to kill the pest kill the parasite or predator. Often the host population is so reduced that the beneficial insects no longer have food and therefore leave. In their absence, unless spraying is continued, the host population can build up (and become resistant) before the beneficial insects can repopulate the area.

The third crisis brought on by insecticides arose from a combination of chemical and ecological factors. The chemical factor was the persistence of many compounds. DDT is the classical example. This compound, like much of the plastic trash that accumulates in our dumps, is not easily "biodegradable"; that is, organisms and processes that normally reduce complex materials to simple elements that can be returned to the soil and reused have little or no effect on DDT, so that it remains as it is, as toxic as ever. It is estimated that there is the equivalent of 0.25 pounds of DDT per acre of arable land in the whole world. The atmosphere and the oceans also contain measurable amounts. DDT applied on land finds its way into rivers and eventually accumulates in the oceans. Indicative of the rate at which the oceans are being contaminated is the revelation that two shipments of canned jack mackerel caught off the coast of Southern California in July 1969 were intercepted by Food and Drug Administration inspectors in New York because they contained residues of DDT in excess of 10 parts per million, the allowable maximum level being five parts per million. It takes somewhat less than ten years for DDT to begin to break down, and little is known about the environmental hazards of DDE, the first product produced in the step-wise breakdown. And DDT is not alone.

The ecological factor with which this persistence combines is the food-chain. To understand these relationships we must remember that while the supplies of energy coming to earth from the sun are practically limitless, the supplies of elements, carbon, nitrogen, oxygen, phosphorus, etc. on earth are finite. All living creatures are just temporary combinations of the stuff of which the world is made. For a time the material of the earth becomes an organism, the environment flows through the organism in the form of food, water, oxygen; it is incorporated into the organism, becomes part of it for awhile and part of the organism is detached (dead tissue, carbon dioxide, urine, feces, etc.) and returns to the earth. When the organism dies, it all returns to the earth. In the process, however, one organism feeds another. Grass extracts nitrogen, water, and minerals from the soil, and oxygen and carbon dioxide from the air; grasshoppers eat the plant; sparrows and field mice eat the grasshoppers; snakes and foxes eat the mice; hawks eat the sparrows. When the hawks and foxes die, carrion flies, carrion beetles, bacteria, and others eat the carcasses. Some of the basic elements are returned to the soil where new generations of plants begin the cycle again. The flies and beetles, becoming adults, are eaten by birds and mice, so on *ad infinitum*. If there is DDT on the plants or in the grasshoppers, the mice and sparrows accumulate large amounts of it through their diet. In effect, they concentrate it. As DDT is passed up the food chain from plant to hawk or fox, it is concentrated even more. The combination of persistence and concentration in the food chain results in the disasters popularized by Rachel Carson.

The three crises engendered by the cavalier rise of insecticides represent the debit side of the ledger, and it is indeed a catastrophic debit. As we have shown, warning voices were raised from the onset. They were drowned by the cheers of insecticide champions. Few had the wisdom or influence to utilize the biological and ecological knowledge to anticipate the problems. As one contemporary writer observed, "Much basic information already known to science is not finding its way into the decision-making process (*Bio Science*, 1970, 20-24)." It took Rachel Carson's *Silent Spring* to awaken people. Read the reviews of this book. It was treated with condescension, arrogance, and even scorn by many professionals. As Doutt and Smith (8) pointed out, "The stupid and simplistic question 'Bugs or people?' was symptomatic of the limited thinking of those afflicted by the pesticide syndrome." Rachel Carson and those who believed as she did were scorned by the chemical industry as "bee, bird, and bunny lovers." *Silent Spring* was dismissed by the professionals as an "emotional appeal." Indeed it was, but nothing compared to the emotional pitch for so long the trademark of the insecticide industry—"Spray or Starve!"

THE BUSINESS BLOCKADE

Insecticide manufacture is big business. Understandably the basic motivation is profit. The more insecticides that can be sold, the bigger the profit. As with any business, aggressive merchandising is a necessary element for staying in business and any opposition threatening business must be crushed. The crises engendered by insecticides have put industry on the defensive. An article in the July 11, 1971, edition of the *New York Times* illustrates only too well the profit motive of insecticide manufacturers, the continued appeal to emotion, and the attitude that well-meaning and innocent industries are being maligned and unjustly chastized. The article, entitled "Ecology: Pesticide Makers' Bête Noire," reports that companies are quitting the field because there is no longer money to be made; some are "tired of justifying their efforts to ecological critics"; others "fear that the proposed Federal Environmental Pesticide Control Act would become too burdensome."

In 1964, sales were $218 million; in 1967, they were $639 million; in 1970, $722 million. In the United States alone, 350 million pounds of insecticide are sold annually. A survey by the National Agricultural Chemical Association among companies that account for 81 percent of total pesticide sales showed that the costs for research and development between 1967 and 1970 rose from $50.4 million to $69.9 million. During the same period, expenditures on safety and environmental testing rose from $7 million to $16 million; however, ecology cannot be blamed for all of the increases because just for economic reasons the cost of discovering a new pesticide rose from $3.5 to $5.5 million and the development time lengthened from 60 to 77 months.

It is all too clear that insecticides are not manufactured to meet a biological need and the growers' demand, but that, as elsewhere in our economic system, a demand is created. Some of the techniques employed in creating a demand are described in an angry article in the *Bangor News* by Stewart Udall and Jeff Stansbury:

Fanned by ignorance and the propaganda of Union Carbide and the U.S. Agriculture Department (USDA), a flash fire of hysteria is now crackling through the Middle Atlantic states. Its bugaboo is a voracious but vastly overrated insect called the gypsy moth. The favorite tactic (borrowed from Union Carbide) is to shower local newspaper editors and politicians with photographs of defoliated trees, creating the false impression that gypsy moths kill most of the trees they invade. Even the *Washington Post* recently fell for the ruse. It printed a USDA photo of leafless trees on Cape Cod with a caption stating "Gypsy moth caterpillars do this to a forest." The same trees were in fresh bloom a few weeks later.

Sevin also persists for far longer than Union Carbide would have us believe. The company's literature on this topic is fascinating. During the late 1950s and early 1960s, Union Carbide highlighted Sevin's "long-lasting residual effects." Then, as the public grew alarmed over persistent pesticides, the company began talking about how quickly Sevin dissipated in the environment.

This, then, is the lethal agent that USDA and Union Carbide hope to spray through the Middle Atlantic region next year. If they can persuade enough journalists and public officials that the gypsy moth is a real devil, the helicopters and airplanes will take off in the spring and dump thousands of pounds of Sevin on woods, wildlife and people.

Wittingly or otherwise, the newspapers have indeed helped to instill panic in the population. The *Evening Bulletin* (Philadelphia) of March 7, 1972, carried a lead article entitled "Gypsy Moth Damage Expected to be Worse." It reported proposed spraying of Sevin on 20,000 acres of "high use" forested and recreational and residential sections of Pennsylvania. The USDA has been taken to task for its role in these happenings by the Environmental Defense Fund and the National Environmental Policy Act (*Science* 1 October, 1971).

What can be said on the credit side of the ledger? Books, pamphlets, and scientific papers devoted to control by insecticides are replete with pictures of sprayed and unsprayed plots side by side, the one ragged, spotty, brown, and bare, the other full, lush, and green. Anyone can see that the insecticide has given beautiful control. The yield is high, the product unblemished, the quality excellent. The grower makes a profit, the consumer is satisfied. Without the insecticide the yield would drop, as would also quality and consumer acceptability. In some instances the crop could not be grown at all. At the same time, profit and acceptability today must be balanced against profit tomorrow and the next day and the next. It must be balanced against other features in the economy. Just as crop failure must be measured in terms of its wider effect on all the ancillary and peripheral economies, so the profit to the farmer must be balanced against the cost to the rest of the community. All too commonly, loss to the farmer is calculated in terms of the wider socioeconomic costs imposed by use of control measures required to maintain profit.

At the same time, the consumer must share part of the blame for the pesticide syndrome. He will not buy blemished apples even though they may be just as delicious as unblemished ones. No fruit or vegetable showing insect damage or, heaven forbid, having a live insect on it would be tolerated. Other agencies are also at work establishing unrealistic standards of quality. The U.S. Food and Drug Administration will not permit a single live aphid to accompany a head of lettuce to market. In the field of processing, the prepack trade has imposed a zero tolerance for pest damage even though 1-2 percent is tolerable for other uses. The frozen-food industry made lygus bugs on lima beans a pest simply because the beans must be near perfect in appearance. Previously, the slight brown blotches appearing occasionally on beans were completely unimportant. Some executive simply decreed that beans should look perfect, and ergo! a pest was created. *All* lygus bugs had to be killed. People have been educated to expect flawless produce. The question is, are they willing to pay the environmental and toxicological price?

One can not have it both ways. Unless people can be re-educated in this respect and as long as agriculture can not be divorced from economics, it is absolutely certain that the use of insecticides must be continued. The conditions under which crops are grown, the artificiality of the whole ecosystem of farming, the slowness of the evolutionary process, and *the irreversibility of the changes that man has introduced throughout the world* are such that the insect-plant relationship evolved over millions of years is now being kept by man in an artificial state of imbalance. This imbalance can be maintained only by man's continued interference. The system can not be left to itself, to the balance of nature. The basic question is: How and what level of control is to be achieved?

To advocate the total abolition of insecticides is, unfortunately, quite unrealistic. They remain a most powerful tool, one that can operate economically and effectively over a wide area, and one that can be marshalled quickly in any emergency. The issue is one of value. It is stated well by one economic entomologist, R. L. Zimdahl, who admits to being a salesman for the technological utility of pesticides:

None of us can deny the efficacy or utility of many pesticides. These are well defined and accepted, even by most critics. However, it does little good to point out solely the advantages. We all acknowledge them and constantly remind ourselves of them and thereby strengthen our convictions. Such mental exercise becomes superficial and may ignore the basic value question raised by the use of pesticides which may impair human health. Value questions are being raised by many responsible people who do not a priori question the need for pesticides as tools of modern agriculture but who are concerned about their use and long-term effects.

Under what conditions, in this country, is it possible to say that there is any pesticide so necessary that any risk of one of these effects is acceptable? *My family would have to be in a more than highly theoretical danger of starving before I would allow any risk* [italics are mine].

To continue to pursue the policy of the past three decades is to court economic and environmental disaster; yet the pesticide syndrome is so firmly entrenched, especially in the minds of the public and those in power, that they barge ahead in the face of all past experience and common sense. The attitude is all too painfully apparent in the measures advocated for the gypsy moth as just described and also for control of the imported fire ant (the United States has three native species).

The fire ant newcomer, *Solenopsis saevissima*, a native of South America, was first noticed in Alabama in 1918. It is now found in nine southern states, from Florida to Texas, where it has taken up residence principally in pasture lands. It stings, as do many of our native ants, bees, and wasps, but this ant is now considered to be a nuisance. The United States Department of Agriculture began an eradication or control program two decades ago. A massive spraying of heptachlor and dieldrin was inaugurated in the 1950s and was a spectacular failure. Beginning in 1962, a new chlorinated hydrocarbon, Mirex, was substituted, with the original goal of eradicating the ant. The Department of Agriculture contemplated a massive program in which Mirex would be applied as an air spray for twelve years over an area of 126 million acres at an estimated cost of $200 million. Despite everything that is known about the effects of persistent chemicals in food chains, resistance, and the dangers of indiscriminate broadcast spraying, the Department of Agriculture began spraying in August. At this point, the Environmental Defense Fund filed a complaint against the USDA. The USDA stopped spraying in mid-November but announced that it would resume in March, deciding to go ahead with its program despite the questionable nuisance value of the ant as a justification for such radical treatment (two documented deaths are known, as compared to the several hundreds by bees and wasps each year) and the known harmfulness of the proposed program. The USDA was taken to court.

A similar attitude bordering on paranoia prevails with respect to the gypsy moth. During the 1950s and 1960s, New England went on a pesticide binge in which it sprayed vast areas from the air with DDT and a newer chemical Sevin. The results were no more successful that the control efforts made decades earlier. In 1971, forestry officials in Virginia agitated for a spray campaign throughout the Shenandoah Valley. Sevin does not control either; it will not prevent future outbreaks and it kills predators, parasites, mammals, and birds. State and federal money is being poured into the battle, the backbone of which is spraying with Sevin. Any town or organized group of local residents can spray to their hearts' content, despite the fact that defoliation by the gypsy moth, while unsightly and a nuisance for awhile, is not an ecological disaster. As one member of the United States Forest Service remarked, "I've called it a cosmetic problem. There may be a better word for it, but it's not digging into the economic base." Normal mortality of the trees involved is about 2 percent. The average mortality of oaks following 90 percent defoliation by the gypsy moth is 5 percent. In some cases the mortality of white oaks in the year following defoliation is about 50 percent, but this must be measured against the lack of mortality in the years following; that is, the trees that succumb tend to be those that were due to die soon, anyway. In other cases, white oaks survived five successive defoliations. Eastern hemlock can be hard hit and suffer up to 92 percent mortality; white pine suffers 11-42 percent mortality. Extensive studies of the relation between defoliation and mortality in forests of Connecticut from 1959-70 have shown that the death of major species in twice defoliated stands was only 2.3 percent annually as compared with 1.3 percent in undefoliated stands (23).

The aggression of the insecticide interests continue even at this moment. There is nearly a total ban against further use of DDT; however, other insecticides with the same properties, high toxicity, persistence, broad killing, and ability to concentrate in food chains are being vigorously exploited. They include Mirex, chlordane, heptachlor, aldrin, and dieldrin. A protracted series of hearings in which the Environmental Defense Fund and the Shell Chemical Company, sole manufacturer of aldrin and dieldrin, are facing each other is currently in progress. At stake for Shell is a gross income of about $10 million per year. The question of the *necessity* for retaining these components for crop protection will hardly figure at all in the battle. The profit motive looms very large. One can only wonder at the motivation and wisdom of those who press for these and similar programs and bring down a new plague upon the heads of the gullible public.

Where is the farmer to turn? He is concerned with making a profit. He is receptive to advice; he wants results. He knows about resistant varieties, the need to fertilize, the need to rotate crops, the need to control pests, but he is acutely sensitive to costs. Consequently, he has limited freedom of choice and is reluctant to

Pl. 1. Part of a swarm of Desert Locusts *(Schistocerca gregaria)* which covered 400 square miles in Ethiopia in October, 1958 (Photograph by C. Ashall).

Pl. 2. Brachonid (a parasitic wasp) about to lay eggs in a wood-boring beetle larva under the bark of a dead acacia (Photograph by Dr. Edward S. Ross).

Pl. 3. A brown lacewing (*Sympherobius angustus*) feeding on aphids (Photograph by Dr. Edward S. Ross).

Pl. 4. A ladybird beetle (*Hippodamia convergens*) feeding on aphids. At the left are some of the beetle's eggs (Photograph by Dr. Edward S. Ross).

Pl. 5. Robberfly and a grasshopper it has captured (Photograph by Dr. Edward S. Ross).

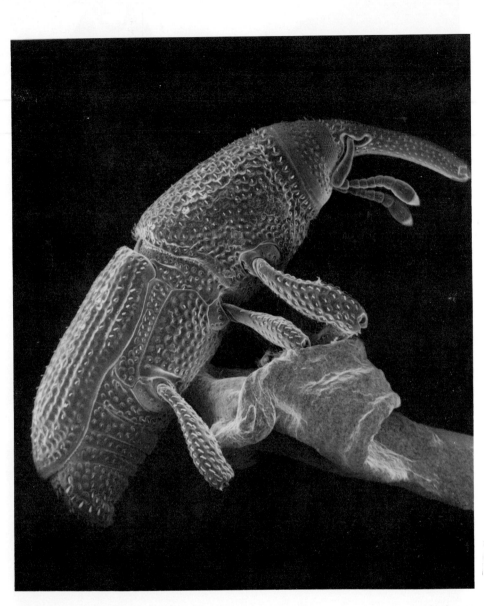

Pl. 10. Rice weevil *(Calandra oryzae)*, a pest of stored grain, magnified 150 times (Courtesy of Dr. C. Pitts).

limit himself even more by techniques that are restrictive. In this view, insecticides strike him as being an efficient and efficacious means of control. How is he to be convinced that he has been receiving bad advice?

Until troubles began to appear, practically all of the advice and all of the control was in the hands of amateurs. Somewhere in the process entomologists have missed the boat. In an introduction to a conference on pest management (18), R. L. Rabb of North Carolina State University wrote: "The failure of the entomological profession to establish reasonable economic thresholds and the reticence of farmers to accept those available have led to many problems associated with the use and misuse of insecticides." And in the same conference (20), D. Pimentel of Cornell said:

I want to make one plea to the entomologists. We're in real trouble with the public, with our students, and with our science colleagues, as far as being responsible scientists. Unless we take a realistic and objective view relative to pesticides and how they are to be handled, we will not gain the respect of these groups. I would like to plead that our Entomological Society, which is in part sponsoring or at least condoning this conference, become active and advise the public which pesticides are essential and which pesticides are not essential and how we're to proceed. And unless we do, we're going to have decisions forced upon us whether we like them or not.

The fact that this view was not shared by others indicates the disarray within the entomological fraternity with respect to this role. There was a feeling among some that leadership should come from universities and individuals at experiment stations. Others have suggested that there should be a well-trained and licensed corps of advisers separated by law from profit and sales in much the way advice of physicians is separated from the sale and profit of dangerous drugs. B. P. Beirne (20) of Canada argued:

The key phase is this: governments must be sufficiently influenced. Why? Because management of people is basic to the management of pests and only politicians and government administrators have or can get the financial and legal powers to manage people appropriately. Who can influence governments? The public. Why should the public do so? Because of the realization of the consequences to them of the intensifying deterioration of the environment. Who can best guide the public in this? We can. In fact, it is our moral responsibility as pestologists to do so. If we do not, nobody else will or can. Moreover, the results of such activities are, I suggest, an essential primary requisite for the development and application of feasible and practical pest management systems.

The political overtones were brought out by P. W. Geier (20) of Australia in answer to the question: Why are some insect pests deemed worthy of sustained attention while others are not?

The answer is surely not because of what they cost, since no comparative relation can be established objectively between the significance for the national economy of say, the housefly and that of aphids. What matters here is *subjective awareness*, the degree of awareness that exists of a given pest species, at a given time.

How is this awareness achieved? Not through any computation of cost, which is a secondary matter. But rather through the fact that the pest problem relates to a situation, or to an enterprise, of sufficient *appeal* to seduce the minds of those in a position to "buy it": the trend-setters and decision-makers of all ilk, and, ultimately, the shareholders, the taxpayers, and the voters at large. A possibility, or a scheme, which possess this appeal becomes what we shall term a *political asset*.

References

1. Anon., Fire ant control under fire. Science.
2. Brown, A. W. A., Insecticide resistance comes of age. Bull. Ent. Soc. Amer., 14 (1968), 3-9.
3. Carson, R., *Silent Spring*, Houghton Mifflin Co., Boston, 1962, 368 pp.
4. Carter, L., Pesticides: environmentalists seek new victory in a frustrating war. Science, 181 (1973), 143-45.
5. Crow, J. F., Genetics of insect resistance to chemicals. Ann. Rev. Ent., 2 (1957), 227-64.
6. Cushing, E. C., *History of Entomology in World War II*. Smithsonian Institute, Washington, D.C., 1957, 117 pp.
7. Dethier, V. G., To Each His Taste. Bull. Ent. Soc. Amer., 14 (1968), 10-14.
8. Doutt, R. L., and Smith, R. L., The Pesticide syndrome-diagnosis and suggested prophylaxis. In: *Biological Control* (C. B. Huffaker, ed.). Plenum Press, New York-London, 1971, pp. 3-15.
9. Entomological Society of Canada, Pesticides and the Environment. Brief prepared at the request of the Board of Governors—1970. Suppl. Bull. Ent. Soc. Canada, 3 (1970), 1-16.

10. Essig., E. O., *A History of Entomology*. MacMillan, New York, 1931, 1029 pp.
11. Gunther, F. A., and Jeppson, L. R., *Modern Insecticides and World Food Production*. Chapman and Hall, London, 1960, 284 pp.
12. Harris, T. W., *Report on the Insects of Massachusetts Injurious to Vegetation*. Fulsom, Wells, and Thurston, Cambridge, Mass., 1841, 459 pp.
13. Hocking, B., Profit without honour. Quaestiones Entomomologicae, 6 (1970), 1-2.
14. Hoskins, W. M., and Gordon, H. T., Arthropod resistance to chemicals. Ann. Rev. Ent., 1 (1956), 89-122.
15. Howard, L. O., *A History of Applied Entomology*. Smithsonian Misc. Pub. Washington, 84 (1930), 1-564.
16. Huffaker, C. B. (ed.), *Biological Control*. Plenum Press, New York, London, 1971, 511 pp.
17. Melander, A. L., Can insects become resistant to sprays? J. Econ. Ent., 7 (1914), 162-72.
18. Osborn, H., *A Brief History of Entomology, Including Time of Demosthenes and Aristotle to Modern Times*, Spahr and Glenn, Columbus, Ohio, 1952.
19. Popham, W. L., and Hall, D. G., Insect eradication programs. Ann. Rev. Ent., 3 (1958), 335-54.
20. Rabb, R. L., and Guthrie, F. E. (eds.), *Concepts of Pest Management*. North Carolina State University Press, Raleigh, 1970, 242 pp.
21. Riley, C. V., Insect Life, vols. 1-7, 1888-98, Washington, D.C.
22. Ripper, W. E., Effect of insecticides on balance of arthropod population. Ann. Rev. Ent., 1 (1956), 403-38.
23. Stephens, G. R., and Hill, D. E., Drainage, drought, defoliation, and death in unmanaged Connecticut forests. Bull. Conn. Ag. Exp. Sta., 718 (1971), 1-50.
24. Strickland, E. H., Could the widespread use of DDT be a disaster? Ent. News, 56 (1945), 85-88.
25. Udall, S., and Stansbury, J., Gypsy moth defoliation not danger to economy. Bangor Daily News, 1971.
26. Usinger, R. L., The role of Linnaeus in the advancement of entomology. Ann. Rev. Ent., 9 (1964), 1-16.
27. Weed, C., *Insects and Insecticides*. Published by C. M. Weed, Hanover, New Hampshire, 1891, 281 pp.

28. Weiss, H. B., *The Pioneer Century of American Entomology*. Published by H. B. Weiss, New Brunswick, 1936, 320 pp.
29. Wilcke, G., Ecology: Pesticide Makers' Bête Noire. The New York Times, Sunday, July 11, 1971, p. 2f.
30. Wood, B. J., Integrated control: a critical review of case histories in developing countries. Address at 14th International Congr. Entom., Canberra, Australia, August 28, 1972.
31. Zimdahl, R. L., Pesticides—a value question. Bull. Ent. Soc. Amer., 18 (1972), 109-10.

Chapter Five
God's Cows

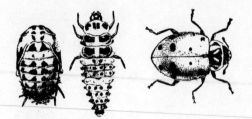

5
God's Cows

Centuries ago, nobody is quite certain when, it was discovered that ladybird beetles were able to keep plants free of aphids. Presumably because of their beneficial ministrations, they earned in France the soubriquets "Vaches à Dieu" (God's Cows), "Bêtes de la Vierge" (the Virgin's cattle), and in England, "Our Lady's Birds." It was reported in 1890 by A. Sidney Oliff, Entomologist of New South Wales, with respect to ladybird beetles that:

In the hop-growing districts of the south of England these swarms occasionally occur, and I have myself seen them in such numbers that they had to be swept from the pathways about the houses. In seasons of scarcity, women and children collect the ladybirds in certain parts of Kent and sell them to the hop grower, who afterwards sets them free, a practical application of one of nature's benefits, which, as far as I am aware, is almost unique in the history of economic entomology, but one, nevertheless, that has prevailed for many years, if not for centuries.

This is not, however, the earliest recorded use of predaceous insects to control herbivorous species. Since very ancient times, Chinese citrus growers have purchased nests of predaceous ants and placed them in the mandarin orange trees, and, now as well as then, they assist the ants in traveling from one tree to another by erecting bridges of bamboo rods. The date growers in Yemen employ the same strategy by placing in the palm nests of predaceous ants collected in the mountains.

The idea that insects injurious to plants can be controlled by predatory insects is thus a very old one. There is a brutal straightforwardness about predation. It is something that man has understood since the beginning of time! As an astute observer of the world around him, man could hardly have failed to notice that some insects preyed upon others. He did not have to understand the finer biological characteristics of predation in order to put it to use; however, with a more intensive knowledge of the phenomenon, he has been able to apply it to agricultural needs in a sophisticated and effective manner. Furthermore, as his knowledge of the widespread occurrence of predation broadened, the choice of suitable predators has correspondingly expanded. Hunters of insects are, after all, found in all groups of animals: crustacea, fish, toads, frogs, snakes, lizards, birds, mammals, and insects. All are potential allies of man, but the hunting skills of vertebrates have seldom been exploited. One outstanding exception is the employment of fish to control breeding of mosquitoes.

PREDATORY INSECTS

Insects themselves are their own worst enemies (a situation that mirrors the relations among men). They provide a large inventory from which to select agents for biological control. There are those that rely upon agility, strength, speed, and ferocity; there are others, equally efficient, that rely upon stealth, camouflage, patience, and duplicity. The speedy ones are represented by many species of ants (the wolves of the insect world), the hawking dragonflies (one caught and examined by an entomologist had crammed more than one hundred mosquitoes into its mouth and could no longer close its jaws), roly-poly ladybird beetles, shiny iridescent ground beetles, and many venomous wasps. The slow, wily predators include praying mantids, squat ant-lions, improbable-looking larvae of swift tiger beetles, and blind, fumbling larvae of flower flies.

Some of the structural and behavioral modifications by which the killing effectiveness of insect predators is perfected are bizarre indeed. Ant-lions, larvae destined to metamorphose into delicate lace-wing flies (Neuroptera), are endowed with unusual jaws and unusual excavating abilities. They construct funnel-shaped pits in patches of fine sand or dust, funnels with sides so steep that the least disturbance causes multiple avalanches. Each ant-lion lies concealed in the earth at the bottom of his pit. When, as eventually happens, some wandering insect or spider tumbles into the pit, the ant-lion instantly flicks up clouds of sand so that the intended victim has his footing so completely undermined that he slides fatefully to the bottom. His arrival is greeted with a lightning thrust from a pair of needle-sharp mandibles which channel a flow of corrosive digestive juice into the wound. In short order the victim is sucked dry. A few flicks of the ant-lion's head and the shriveled corpse is tossed out of the pit, and the trap is reset.

Another stealthy fellow is the larva of the tiger beetle. He too lies concealed in the ground, snug in a vertical tunnel, the trap-door of which is his flattened head. This is equipped with a formidable pair of piercing mandibles. As added insurance for success, the larva sports on its back near its tail two stout curved spines which anchor it to its burrow. There is no possibility for the intended victim to drag his attacker from his lair.

The maggot-like larva of the flower fly hardly exhibits the mien of a killer as it crawls blindly over the surfaces of leaves; however, its prey is not so agile, either. When the larva bumps into an aphid, it immediately buries its hooked mouth in the soft body, hoists the aphid into the air as Hercules lifted Atlas, sucks it dry, and flicks away the empty shell.

Perhaps the most bizarre behavior is that of a Javanese bug (*Ptilocerus ochroceus*), described by Brues as follows:

The *Ptilocerus* bears on the underside of its abdomen a tuft of beautiful fiery red trichomes which serve to conceal the opening of a gland and to disseminate its secretion which is fed upon by a common ant of that region. . . .After some manoeuvers on the part of the bug, the attention of the ant is attracted to the organ which is licked energetically by the unsuspecting victim. After the ant has imbibed this exudate from the surface of the trichomes for a few minutes, it becomes apparent that a dose of "knock-out" drops has been included in the potion; the ant staggers and falls. The wily *Ptilocerus* now grasps her with its forelegs and thrusts its beak through a weak spot in the back of her neck. Without further ado, the unconscious dupe is quickly sucked dry, its empty shell cast aside and *Ptilocerus* is ready to entice another victim.

Two characteristics of predators fit them well for the role of control agents: their insatiable appetites and the specificity of their diets. Their rapacity is incredible. An aphis-lion may consume thirteen aphids daily; another requires 500 eggs of the oriental fruit moth to reach maturity; the larva of a ladybird beetle may consume 474 aphids at the rate of 24 each day, followed in adulthood by 34 each day; another may require 772 as a larva and 791 as an adult. The large Calosoma beetle, imported to combat gypsy moths, requires 41 caterpillars to become full-grown. The larva of a syrphid fly may consume 474 aphids during its development.

Contrary to general opinion, predators are not indiscriminate hunters. Most insect predators have become so highly specialized structurally that the great efficiency they have achieved by becoming specialists has been partly offset by the restrictions it imposes on the kinds of prey that they are capable of attacking. The advantage of this characteristic for biological control is that specific predators can be employed to combat specific pests without endangering beneficial insects.

The idea that parasites could serve the farmer in the same role as do predators developed more slowly because the nature and extent of parasitism was not fully appreciated until the beginning of the eighteenth century. One of the earliest descriptions of parasitism appreciated as such was that of the Dutch microscopist Antonie van Leeuwenhoek, who illustrated and discussed a sawfly parasite in 1701.

Although nearly all orders of insects boast some parasitic members, most parasites are wasps of the superfamilies Ichneumonidae, Chalcididae, and Syrphidae, and flies of the family Tachinidae. Only the immature stages are parasitic. Females lay their eggs on the host's body, its egg, or its food. The smallest of parasites lay their eggs on the egg of another insect, and the young parasite completes its entire development on the contents of that single egg. Parasites that lay their eggs directly on the host most frequently attack immature stages. Thus, the larvae of ichneumon wasps and tachnid flies develop within the bodies of caterpillars, where they feast so skillfully on the least critical tissues that the host may complete its larval development and even pupate before dying. Many a young naturalist raising caterpillars has seen his lovely creature suddenly sprout a coat of small white cocoons (spun by emerging parasite larvae), die, and in dying give birth to a host of small wasps. Or he has seen his carefully tended chrysalis give rise to a hairy fly instead of a shimmering swallowtail butterfly.

Strictly speaking, microorganisms that cause diseases in insects are also parasites. Appreciation of the nature and potential of microbial infection came even later than the understanding of parasitism, even though awareness of the diseases of domestic insects has a long history. As early as 2700 B.C., the Chinese observed diseases of silkworms and honeybees. In the West, Greek myths relate that Aristaeus, a beekeeper, son of Apollo and Cyrene, lost his hives through disease. The earliest historical account of diseases of insects is Aristotle's description in his *Historia Animalium* of devastating maladies suffered by honeybees. In the Middle Ages, a mysterious malady of silkworms was common knowledge. As with parasitism, however, the biological nature of these diseases was not understood. The first published description of a pathogen was de Réaumur's account in 1726 of a fungus growing from the body of a caterpillar (27, 28). The fungus, *Cordyceps*, grows tree-like from its host. Early observers could not decide whether the worms grew from plants or the plants from worms. In China, these "plant worms" were considered to be rare drugs of great value. (They could still be purchased in San Francisco as recently as 1956.)

One hundred years after de Réaumur's article, an English clergyman, William Kirby, published with Spence "An Introduction to Entomology," in which he wrote a remarkably enlightened chapter on the diseases of insects (27, 28). It was not until 1834, however, that the causal relation between insect diseases and microorganisms was established. Agostino Bassi, the "father" of insect pathology, performed a classical experiment which demonstrated that a microorganism (a fungus) could cause disease in an insect (the silkworm). Not long afterwards (1836), he suggested that the activities of microbial life could be employed to destroy insects. Unfortunately, nobody attempted to capitalize on the idea.

In the meantime, the stage was being set for a devastating demonstration of the ability of disease to decimate populations of insects. The blow fell on the French silk industry. Beginning in the thirteenth century, France had become a leading producer of silk, supplying one-tenth of the world's supply. In the middle of the nineteenth century, plague hit the silkworms, and production fell from 26 million kilograms per year to a mere 4 million. Nearly 4,000 nursery owners turned to the government for help. It in turn persuaded Pasteur to investigate the pestilence. Although a chemist at that stage in his career, he discovered that two diseases were responsible, pébrine and flacherie. He recognized the microbial basis of these diseases and the mode of transmission, discovered how to control pébrine, and in so doing saved the silk industry of France.

The moment seemed ripe to resurrect Bassi's idea of microbial control, but the suggestion, when it did come, came from the other side of the world. An American entomologist, John L. Le-Conte, in an address in Portland, Maine, in 1873 proposed that fungi be employed for control (27, 28). His broad proposal was followed six years later by a very specific proposal by a German-American entomologist, H. A. Hagen. Finally in 1882, Pasteur himself made a similar proposal. But the first significant experiment in which an *injurious* insect, the wheat cockchafer (*Anisoplia austriaca* Hbst.), was deliberately infected with a pathogen with a view to control was that of the Russian, Metchnikoff, in 1879.

Microbes are a motley crew. They include bacteria, viruses, rickettsia, fungi, and protozoa. Flacherie of silkworms and foul-brood of honeybees, two of the most ancient of microbial diseases, are bacterial. The most commonly observed disease of houseflies is a fungus (*Entomophthora muscae* [Cohn]). While it is in the process of killing, it hardly disturbs the victims. Their corpses are frequently observed clinging in life-like positions on window-panes, each surrounded by a halo of spores. Other fungus diseases are the muscardin diseases, so called because the mat of fungus on an infected insect causes it to resemble a French bonbon (muscardin). One genus (*Beauveria*) of this group is especially destructive to chinch bug, codling moth, and European corn borer. Death may ensue within three days. A mummified corpse remains, packed with infectious conidia (spores) which under favorable conditions can be disseminated far and wide.

Many protozoa infest insects, the most celebrated being *Nosema bombycis* Naegeli, the pébrine of silkworms. The number of virus infections is also large. Some viruses (the polyhedroses) cause polyhedron-shaped bodies to be formed in the nuclei of the host's cells; others in the cytoplasm. Other viruses (granuloses) cause granular bodies to form in the host's cells. Once a virus has invaded an insect, most often larvae of moths and butterflies, the body almost literally melts as the infection develops. Complete disintegration rapidly follows.

Predators, parasites, pathogens! For millions of years they have kept the insect hordes in check, the predators striking quickly with forthrightness, the parasites more subtly, more slowly and selectively, and the pathogens with devastating rapidity. A few imaginative men had visions of how these powerful natural forces could be recruited in man's service. Unfortunately, much of the basic knowledge necessary to turn their ideas into action—knowledge of the habits of insects, of ecology, behavior, and epidemiology—was lacking, and, as a result, most of the earliest attempts at organized biological control were failures. The necessary knowledge and experience could be acquired only if more qualified people could be enticed into the field of biological control and only if influential administrators could be persuaded of the wisdom of that course. Either a crisis or smashing success was required. Both came in the New World, and first honors fell to the predators.

Many thoughtful entomologists viewing the state of agriculture in America became concerned with the increasingly heavy impact of injurious insects. It is doubtful that they appreciated all of the complex causes underlying the trouble. One of the more obvious, the introduction of foreign insects into an environment where they were freed from attacks by their natural parasites and predators, was appreciated by two men. One, Asa Fitch, State Entomologist for New York, was tempted to speculate on this matter when he discovered that a wheat midge particularly destructive of wheat in America was a native European insect, but one that was not a serious pest in its own land. He asked in 1856: "What can that cause be? I can impute it to only one thing. We are here destitute of nature's appointed means for repressing and subduing this insect . . . .We have received the evil without the remedy."

The solution was obvious to Fitch, who then attempted unsuccessfully to import some parasites from Europe. His lack of success was caused by administrative rather than biological difficulties. In public, private, and governmental circles there was great apathy with regard to the importation of parasites. Even at this early date, there was little general interest in the idea of combatting insects with insects, presumably because the action of beneficial insects was unobtrusive and slow compared to the insecticides even then available. Furthermore, the phenomena underlying the concept of biological control, namely, population ecology, the "balance of nature," the "web of life," were not widely understood.

De Réamur (1683-1757) had produced a classical study on insect parasitism; the intricate relations between organisms was one of the foundation stones of Darwin's *The Origin of Species*

(1859); and some of the dynamics of populations were brought to public attention by Malthus (1803) and his contemporaries who were concerned with the alarming fertility and consequent overpopulation of mankind. In Europe, another entomologist, Vincent Köllar, had described in great detail various predators and parasites, the manner in which they preyed upon other insects, and how essential they were to the balance of nature. His manuscript so impressed Emperor Francis I of Austria that he commanded its publication.

In America, Benjamin Walsh, a contemporary of Fitch, tried his best to stir up an interest in biological control by badgering, cajoling, criticizing. He wrote:

But the scientific mind is always ahead of the popular mind. Vaccination, gas, the steam engine, the steamboat, the railroad, the electric-telegraph, have all been successively the laughing-stock of the vulgar, and have all by slow degrees fought their way into general adoption. So will it be with the artificial importation of parasitic insects. Our grandchildren will perhaps be the first to reap the benefit of a plan which we ourselves might, just as well as not, adopt at the present day. The simplicity and comparative cheapness of the remedy, but more than anything else the ridicule which attaches, in the popular mind, to the very names of "Bugs" and "Bug hunter," are the principal obstacles to its adoption. Let a man profess to have discovered some new Patent Powder Pimperlimpimp, a single pinch of which being thrown into each corner of a field will kill every bug throughout its whole extent, and people will listen to him with attention and respect [for Pimperlimpimp read DDT, for the discovery of the insecticidal properties of which a Nobel prize was awarded]. But tell them of a simple, common-sense plan, based upon correct scientific principles, to check and keep within reasonable bounds the insect foes of the farmer, and they will laugh you to scorn.

THE LADYBIRD SAGA

Walsh was not far from wrong. He erred principally in reserving his venom for the public and the politicians. He might equally well have indicted some of his colleagues. The attitudes, emotions, self-interests, and intrigues that permeated the realm of biological control even in those early days is dramatically illustrated by the famous Vedalia ladybeetle case described by one of the later participants, L. O. Howard, in his largely autobiographical history of entomology, and, most entertainingly, by Doutt in an article entitled "Vice, Virtue and the Vedalia." This case included such bizarre episodes as an unhappy love affair, political intrigue, and a pair of diamond earrings. The tale reveals some measure of the heat generated by opposing views, but, more importantly, the wisdom that lay behind the thinking of Fitch and Walsh.

The principals were: Charles Valentine Riley, Chief of the Division of Entomology of the United States, probably the first person to introduce *parasites* from one region to another; Albert Koebele, a naturalized German whose beautiful mountings of insect specimens so impressed Riley that he offered him a job in Washington; D. C. Coquillett, a native of Illinois, who had also been employed by Riley; and Frazer Crawford, an Australian entomologist.

When the curtain rose, Koebele and Coquillett were Riley's field agents in California. Coquillett had moved to California to alleviate a tubercular condition. Riley offered a position to Koebele, who gave up the job he was holding only to discover belatedly that Riley could not honor his offer. Riley tried to make amends by appointing Koebele to a temporary field position at $65 per month. In Washington, Rileys's temporary assistant became involved in an unhappy love affair from which he sought relief, requesting a transfer to some far away place. His request was honored more expeditiously than his job offer had been. He was sent to California, where eventually, in 1886, he was assigned to work with Coquillett. The two did not get along well together. Koebele's criticisms of Coquillett, transmitted to Riley, led to the firing of Coquillett (who was later rehired).

During this period, the cottony-cushion scale, a native of Australasia that had been inadvertently imported sometime around 1868, had multiplied to such numbers that it was forcing Californian orchardists to abandon the commercial growing of citrus fruits. After Coquillett had been fired by Riley, he, together with J. W. Wolfskill, the foreman of a ranch in Los Angeles, privately experimented with hydrocyanic gas as a fumigant for scale insects. Riley was peeved, as the following quotation reveals: "My friend, Mr. Coquillett, perfected this gas after his employment by the Department of Agriculture ceased. But it is a general truth that the moment any person or persons become interested in a patent or in any remedy they desire to control from that moment their judgment can no longer be depended on as to the value of other remedies." Despite its rancor, there is more than a modicum of truth in this accusation, which was contained in an address to the Fruit Growers Convention. In the same address he concluded:

I would not hesitate, as United States Entomologist, to send someone there [Australia] with the consent of the Commissioner of Agriculture were the means for the purpose at my command; but unfortunately, the mere suggestion that I wanted $1,500 or $2,000 for such a purpose would be more apt to cause laughter and ridicule on the part of the average committee in Congress than serious and earnest consideration, and the action of the last Congress [a rider on the appropriation bill for the USDA prohibiting foreign trips by employees, presumably to curtail Riley's frequent European junkets] has rendered such work impossible by limiting investigations to the United States.

At this point, Riley was clearly committed in principle to the idea of biological control of the scale. It is difficult to discover whether or not he was very aggressive in trying to squeeze funds out of the government. The California fruit growers, however, were desperate. They kept badgering Washington until, finally, after some political maneuvering, $2,000 was appropriated, ostensibly to pay the expense of an entomologist to represent the State Department at an international exposition in Melbourne. The Californians wanted Coquillett to be that man, but Riley selected Koebele with instructions to find and ship home parasites of the cottony-cushion scale. The story from here on continues to be replete with jealousies, animosities, charges, counter-charges, secrecy, and credit-seeking.

Koebele sailed for Australia on August 25, 1888, for the express purpose of sending back a parasitic fly (*Cryptochaetum icery* [Williston]) which had been discovered by the Australian, Frazer Crawford, in 1886, and found to be an effective control for the cottony-cushion scale. Twelve thousand of the flies were sent to California, where today, their descendants, together with those of flies sent by Crawford a year before Koebele sailed, still help control scale. Coquillett and his associates were receiving credit from California for the success of the parasitic fly while Crawford, forgotten in far-away Australia, fumed.

Meanwhile, Koebele had discovered a ladybird beetle, the Vedalia, feeding on scale. At first the beetle was ignored. Finally, when its potential was realized, three consignments totaling 129 individuals were sent to California between November 1888 and January 24, 1889, where Coquillett began to breed them. They multiplied furiously and ate scales voraciously. Coquillett, together with his friends of the cyanide days, began to distribute the new predators all over the state. By June of 1889, they had released 10,555 individuals. Orchards that had been on the verge of abandonment were reprieved, the scale ceased to be a major pest, and shipments of oranges from Los Angeles County jumped in one year from 700 to 2,000 cars.

In 1892, Koebele sent another species of ladybird (*Novius koebelei*) which was also effective. The total cost of the project was $1,500 (Koebele's travel expenses). Koebele received as a token of appreciation from the Californians a gold watch and his wife, a pair of earrings. California began immediately an extensive research and development program on biological control. The legislature appropriated $5,000 to send an entomologist on another expedition to Australia. They sought the cooperation of the federal government by suggesting that Washington pay the entomologist's salary. Riley turned down their request but was overruled by his superior, the Secretary of the Department of Agriculture. Relations between Washington and California cooled; Riley recalled his two field agents. Coquillett returned to Washington, but Koebele resigned and accepted a lucrative position in Hawaii.

How does one assess the Vedalia project? One entomologist, E. O. Essig, wrote in 1931 in his history of entomology: "The introduction of the Vedalia and other natural enemies of the cottony-cushion scale has not been the means of exterminating the destructive scale in California, but it furnished to the world the first demonstration of effective natural control and was responsible for the biological control of insects now being successfully carried on in many parts of the world." The method has been described as difficult and slow, but it has failed only where those in charge expected too quick and easy returns or where the project has been directed by incompetent men or not been allocated sufficient funds. It failed completely to control scales on pears near San Jose, California, in 1914-15; however, the failure arose from the fact that the trees had been heavily treated with lead arsenate before the Vedalia were released, and the spray residue killed the beetles. Soon after the popularization of DDT there were other serious outbreaks in citrus groves because populations of predators and parasites were decimated. Today both the fly *Cryptochaetum* and the Vedalia continue to keep cottony-cushion scale in check, except when spraying interferes.

Even at the height of its success, the California program was sharply criticized. One of America's most eminent and influential entomologists, L. O. Howard, first Riley's assistant and then his successor, later expressed a bias against California in particular:

There has been a tendency for many years for persons with strange beliefs to migrate to California, largely on account of the climate, and southern California today (1930) is known as the home of all the heterodoxies, and biological control in general: So great an enthusiasm for natural control was aroused in California by the success of the Australian ladybird that the state made apparently no advances in her fight against insects for many years. Mechanical and chemical measures were abandoned. The subject of natural control held the floor. It is safe to say that a large share of the loss through insects suffered by California from 1888 until, let us say, 1898, was due to this prejudiced and badly based policy.

A detailed refutation of Howard's statement is given by Doutt (9). Howard's many statements leave little doubt that he saw biological control and chemical control as opposing philosophies, and it is clear which philosophy he espoused. Speaking on the campaign against the gypsy moth in New England, he wrote:

The stream of poisoned water thrown up with great force from the powerful machines breaks into the requisite spray long before it reaches the tops of tall trees. All of the features of the machines and of the hose were greatly improved, and it has of late been one of the marvels of applied entomology to see a spraying machine by the roadside in the mountainous regions of southern New Hampshire getting its supply of water from a roadside stream, and through strong sectioned hose carried over the top of hills of considerable size, spraying the trees on the other side of the hill, perhaps nearly a mile away.

OTHER EFFORTS AT BIOLOGICAL CONTROL

Despite the spectacular success of the Vedalia, biological control (the term was not commonly adopted until 1930) got off to a rocky start. Other equally spectacular successes did not follow immediately. There were failures. Introduced predators and parasites did not, for example, decimate the gypsy moth that was defoliating the forests of New England. In Canada, biological control had started in 1882 but languished for lack of interest, successes, and administrative initiatives. Control by pathogens did not fare much better. One of the most publicized efforts in the United States —an attempt to control chinch bugs with a fungus—failed.

In 1865, chinch bugs in Illinois had been observed dying from a white fungus. The fungus was still prevalent in 1888, at which time diseased bugs were collected and deliberately scattered about in hopes of spreading the infection. In the same year, the legislature in the state of Kansas established an Experiment Station to study and produce large quantities of the fungus. As a result, 50,000 packages of fungus were distributed to farmers. In 1891 and 1892, there were encouraging reductions in the population of bugs; but in the years that followed, artificial introduction of the fungus failed to reduce the population as well as natural infestations had in previous years. Scientists concluded that environmental factors were of paramount importance in influencing the effectiveness of fungi in the field and that there was little hope for fungi being reliable means of control since man cannot control the weather. The *Beauveria* project was abandoned in Kansas, Illinois, Nebraska, Missouri, Ohio, and Oklahoma. The year 1900 saw the high-water mark of hope for microbial control. By 1930, skepticism and discouragement had replaced enthusiasm.

After the failure of the chinch bug project, a brief resurgence of excitement occurred during the years 1911 and 1914. In Yucatan, Mexico, d'Herelle had noticed epidemics among locusts migrating from Guatemala. By 1912, populations were so reduced in Guatemala that the locusts no longer invaded Mexico. From infected locusts, d'Herelle isolated a small gram-negative rod (a bacterium) which caused dysentery and septicaemia on their hosts. Infected locusts died within 12 to 24 hours after having suffered diarrhea and vomiting. The cadavers turned black. Insects, too, have their Black Death.

Although no overwhelmingly successful control programs sprang from the new reports, and enthusiasm among applied workers and farmers again waned, research workers continued to search for means of putting pathogens and parasites to agricultural use. Another success was needed to overcome apathy and inertia. In Canada, the brown-tail moth provided the stimulus. This European insect (*Nygmia phaeorrhoea* [Don.]) exploded into a severe infestation in eastern Canada early in the 1900s. Between 1913 and 1915, parasites and predators were liberated. From its maximum in 1914, the infestation declined to zero in 1927. Whether or not the decline was due to biological control, as was long maintained, or would have occurred anyway can never be decided. It succeeded nonetheless in injecting new life into biological control. Other apparent successes followed. Altogether, they were instrumental in sparking the establishment of The Natural Control Investigations Laboratory at Fredericton, New Brunswick.

From 1921 to 1940, scientists in the United States Department of Agriculture discovered and became acquainted with a bacterial disease of the Japanese beetle. This milky disease, called after the opaque white appearance that infected insects acquire, is one of several. (The flacherie of silkworms which Pasteur studied was probably one variety.) The disease is caused by gram-positive rods. Grubs of Japanese beetles (*Popillia japonica*, [Newman]) are particularly susceptible to two species, *Bacillus popilliae* (Dutky) and *B. lentimorbus* (Dutky). These bacteria became a classical example of microbial control.

Between 1939 and 1953, 109 tons of cultured spore powder were distributed to more than 90,000 sites in thirteen eastern states as part of a governmental control program. A typical case was that of a Maryland golf club on the fairways of which there were 20-60 grubs per square foot of turf. The bacteria reduced the population to 1-3 grubs per square foot.

Industry was slow, however, to take an active interest in pathogens. Not until the 1950s were milky disease bacteria produced commercially. It is interesting that householders having trouble with grubs in their lawns were still (e.g., *Philadelphia Bulletin* September 30, 1973) being advised to use insecticides; no mention was made of commercially available bacteria.

Another successful bacterium is the *Bacillus thuringiensis*, first isolated in 1911 by Berliner from larvae of the Mediterranean flour moth in flour from a mill in Thuringia, Germany. This bacterium produces in the bodies of its hosts crystals so toxic that the hosts die within a week. The crystals, composed of protein, survive the death of the host and apparently persist indefinitely. Commercial production of this pathogen began in 1958.

In general, until very recently, no aspect of biological control received much world-wide encouragement. In America (except for the state of California), no sustained major efforts were made, and the funds and manpower allocated were modest. That industry took any interest at all in pathogens was due largely to the efforts of the late Edward Steinhaus in California. His experiments, writings, and energetic proselytizing gradually stirred interest. By 1945, the University of California had established a Laboratory of Insect Pathology. A year later, the Canadian Department of Agriculture followed with a similar laboratory.

THE FUTURE OF BIOLOGICAL CONTROL

Where then does biological control (including microbial control) stand today and what is its potential? Biological control is readily understood, but its definition is elusive, as one of the leading research workers in this field, Paul DeBach, Professor of Biological Control at the University of California, has explained. It can be defined as "the action of predator, parasites, or pathogens in maintaining another organism's population density at a lower average than would occur in their absence." In this sense it is one aspect of natural control which embraces abiotic as well as biotic agents.

Control refers to the regulation of population numbers. Natural control refers to the maintenance of a more or less fluctuating population with certain upper and lower limits by biotic and abiotic environmental factors more or less permanently. The levels may be high or low (some insects are rare, others are abundant), and the fluctuations may be regular or irregular, great or small. The point is that no species multiplies indefinitely.

Three chacteristics of natural control are especially deserving of emphasis: its fluctuating nature; the fact that the average population level can always be high; and its permanency. Chemical control, by contrast, implies reduction of a population to a stable low level and is temporary unless continued indefinitely. Biological control, even though it describes the biotic as contrasted with the abiotic regulatory factors of nature is usually thought of as "applied natural control." It is the study, importation, augmentation, and conservation of beneficial organisms for regulating the population of other organisms, usually pests. It is based on the premise that predators, parasites, and pathogens do help to some extent in regulating populations and that their introduction to a population that is free of them (having been transported by man to a part of the world where their natural enemies are absent or having, through agropractices, risen to levels in their native land where their balance with their natural parasites is at too high a level to be acceptable to man) will result in a new balance being established at a level more acceptable to man.

There is much disagreement among scientists interested in population dynamics about how effective biotic factors are as determinants. Very few, however, would deny that they have some effect. There is a voluminous literature on the subject of population dynamics. For concise comprehensive treatments, especially in reference to insects, the reader should consult the section on the Ecological Basis of Biological Control in DeBach's book and the book by Clark, Geier, Hughes, and Morris.

The evidence that biological control is sound in principle and does indeed work is impressive. There are, to be sure, spectacular failures, as there are with all methods of control; but a critical evaluation of all the attempts up to 1966 revealed a total of 66 cases of complete control, 88 cases of substantial control, and 71 cases of partial control, giving a total of 225 cases in 63 different countries involving about 110 different species of pests.

The criterion for success is an economic one. Complete success means control over a major pest of a major crop over an extensive area so that treatment with insecticides is rarely, if ever, necessary. Substantial success is that where economic savings are less by reason of the pest or the crop being less important, or the area under cultivation restricted, or occasional supplementary treatment with insecticides being required. Partial success is that in which chemical control is still necessary but intervals between necessary applications are lengthened or outbreaks occur less often. Hawaii and California lead in recorded successes. Success is clearly proportional to the amount of research and exploration that has been undertaken and the vigor with which programs have been pursued.

Since the time of Howard, advocates of pesticides have argued that there have been few successes, these have been limited in scope, and it is an unprofitable business. One of the critics of biological control, the entomologist Roy Hansberry, remarked that the number of papers published in the *Journal of Economic Entomology* during the year 1967-68, in which successful control of pests was reported, totalled 34 and all dealt with pesticides. A rebuttal by Huffaker, Messenger, and DeBach stated in part: "It might be noted that workers in biological control have, regrettably, almost ceased sending manuscripts to this journal and many have dropped their subscriptions, such is its emphasis on pesticides." Hansberry further remarked that 81 cases of biological control for Canada and the United States is a paltry achievement. This is a grossly unfair statement considering the huge sums of money that have been spent in the development of insecticides and the powerful merchandising and high-pressure propaganda that has emanated from the pesticide Establishment.

Although the greatest number of successes with predators and parasites have been against scale insects and mealy bugs, especially those infesting orchards, success is not limited to these. Very good results have been obtained against some forest insects in the northeastern United States and Canada. Sometime prior to 1900, the European spruce sawfly (*Diprion hercyniae*) found its way into Canada. Males are rare—but the majority of females can reproduce without mating. The caterpillar-like larvae feed on old foliage of white, red, and black spruce. For approximately thirty years, they chewed their way unnoticed through the forests of Quebec, New Brunswick, and northern Maine, until discovered in 1930, after they had inflicted considerable damage in an area covering several thousand square miles. Since the larvae generally refuse new foliage, trees recovered from the attacks; however, after six years of repeated attack, trees died. The native predators —birds, small mammals, and insects—had more larvae than they could eat and were unable to make a dent in that food supply.

Between 1934 and 1939, Canadian entomologists introduced a number of parasites from Europe. Seven of these parasites found the New World to their liking. At the same time, unknown to the entomologists, a virulent virus was introduced along with the parasites. Together with the parasites, the virus spread throughout the infested area. By 1945, the outbreak of sawflies was over, and from then on the numbers have remained at a level below which serious defoliation occurs. In seven years, the destructiveness of sawflies was eliminated and, for the remaining eighteen years over which records were maintained, kept below pest level. This is considered an outstanding example of the biological control of an introduced insect that had become a pest. Aerial spraying of insecticide would probably have reduced the pest in one year. It would have been spectacularly successful, but it would have been temporary and much more expensive, requiring repeated application; it would eventually have created resistance and poisoned the environment.

Since the classical success of *Bacillus popilliae*, which reduced the Japanese beetle from a major to a minor pest in the northeastern United States within three years of its introduction, there have been numerous successes with a number of microorganisms. For details, the article by Burges and Hussey in their book on microbial control should be consulted.

When biological control works, it is resoundingly economical. Successful projects in California alone for the period 1923-59 resulted in a net saving of $110,990,973. The work during this period was conducted by the University of California. This figure does not include the savings attributable to biological control of the cottony-cushion scale by the Vedalia beetle. Another balance sheet, especially accurate, is available in the instance of control of coconut scale which threatened the entire economy of the small islands of Principe off the coast of West Africa. The initial, and only, cost of biological control was $10,000. A conservative estimate of the benefits that accrued over the first ten years following control is $2 million.

A final example of economic gain concerns the winter moth (*Operophtera brumata* L.) accidentally introduced into Nova Scotia, Canada. Within ten years, it damaged oak forests in two counties to the extent of about 26,000 cords per annum. The value of the destroyed trees would approximate $2 million in today's market. It is estimated that the potential loss, had the moth not been contained, would have been not less than $12 million. To control the moth, two parasites were introduced from its home in Europe, a wasp and a fly. The total cost of research related to the liberation of the parasites was about $160,000.

THE THREAT OF CHEMICAL CONTROL

It is not the difference of opinion among students of insect demography, or the failures, that have stood in the way of the development and use of biological control. It is the shadow of chemical control and the severe criticism, not unmixed with condescension, that has been directed toward it ever since the ascendency of chemical control as a major industrial and sociological force. Many advocates of chemical control continue to be generally hostile to biological control. As a consequence, biological control has been neglected.

The codling moth on apples may serve as an example. Until 1960, millions of dollars were spent on sprays and on research aimed at developing better insecticides, while virtually nothing was done toward searching for and introducing beneficial insects to prey upon the moths. Only two parasites of the codling moth had been introduced into the United States from among the 120 species known to exist in various parts of the world. In Canada, the record of attempts on apple (and pear) is about the same. In other cases, of 600 or more insects known to be economically important pests, only 92 have had natural enemies imported as agents of control. Were it not for the state of California, the showing would have been even more dismal. That state has been instrumental in procuring two-thirds of the number imported.

It is not difficult to understand the psychological handicap under which biological control labors. The apparent instant success of synthetic organic insecticides after World War II launched a massive program supported by the platform that: they give better protection than other methods (which is true only over the short haul); they have a broad spectrum and so can be used against different kinds of insects (and beneficial insects) on the same crop; they have a residual effect that greatly prolongs their activity (and gets into the food chain); they can be cheaply made (a new insecticide costs $3.4 to $5.5 million to develop); they are easily applied (they require complex spray equipment or aircraft).

Many people continue to be biased toward insecticidal control. The brief of the Entomological Society of Canada contains the following statement:

A prevalent misconception is that there are many alternative methods available for immediate use if pesticides are withdrawn. Unfortunately this is not so, partly because the wide initial success of pesticides reduced the apparent need to strengthen research on other methods [evidence of the great lack of perception on the part of those making decisions], and partly because most alternative methods require a specialized approach, tailored to a particular pest or crop, and are correspondingly more difficult to develop and apply [a weak and fallacious argument]. It must be concluded that: *To achieve present levels of pest control, chemical pesticides are, for the time being at least, essential.*

There is, of course, truth in the conclusion; however, one can question the necessity of maintaining *present* levels of control. Also, unfortunately, the continuing *necessity* delays and impedes development of alternative methods. As long as agriculture is addicted to pesticides and as long as one addiction that has bad side-effects can be substituted for another there is little incentive to search for cures.

Apart from the psychological obstacle to the development of biological control and other nonchemical methods of control, there is the formidable financial handicap. Despite the fact that large sums are not required, it struggles along on a pittance, compared to chemical control, because industry will not pour millions into its development and promotion. There is no profit in it. Only private organizations and governmental agencies provide funds, and these are quite limited. Nor is there much money to be made in microbial control at the moment. Microbes can not be patented (although techniques for their culture and formulation can). *B. thuringiensis* is now produced commercially in California and elsewhere in fermentation tanks of 40,000 liter capacity. In Russia, near Kiev, a factory is being built to produce 250 tons per year of the fungus *Beauveria bassiana*. Viruses (not a "new" class of pesticides as *Science* recently [September 7, 1973] headlined) are also now mass-produced. Nevertheless, these are niggling efforts compared to the ongoing development of chemical insecticides.

Additionally, for pathogens there are legal problems. Microbial products intended for pest control fall under the regulations of the 1947 Federal Insecticide, Fungicide, and Rodenticide Act. Consequently, the Pesticides Regulation Division of the Department of Agriculture and the Food and Drug Administration must clear a pathogen before it can obtain federal registration. As of 1971, only two pathogens, milky spore disease bacteria to control Japanese beetles and products containing spores of *Bacillus thuringiensis*, had been registered.

In 1962, efforts to have a virus registered were initiated by the United States Department of Agriculture, and since 1965 two commercial firms have been attempting to obtain registration for the nuclear polyhedrosis virus (NPV) of *Heliothis zea* (the cotton bollworm). Registration may at long last be near because in the *Federal Register* of May 30, 1973, the Environmental Protection Agency exempted the virus from the requirement of leaving no more than a minimum of residue on crops.

The cause of biological control is also hurt by those who take the extreme position that predators, parasites, and pathogens can completely replace chemical control. The cause is also diminished by claiming success on the basis of incomplete data or poorly controlled field tests. Beirne in Canada has shown, for example, that the alleged success of biological control against the European stem sawfly in Ontario might in fact be a case of failure, because much of what happened could be attributed to normal fluctuations in populations, to changes in methods of harvesting, and to idiosyncrasies in census taking.

As we have repeatedly pointed out, the balance of nature in agriculture is really an artificial imbalance which man, for economic reasons, is endeavoring to maintain in favor of the plant over the insect. Biological control alone can not redress the evolutionary and ecological disturbances that man has engineered. Artificial control in the form of chemicals must be included if the status quo is to be maintained.

It is clear to the more sober specialists that some pesticides have to be used judiciously, intelligently, carefully. A total ban on all insecticides is unrealistic now and possibly ever. When insecticides dominated agriculture, other existing control measures were abandoned, among them many sound agricultural practices. No longer were crop remains that offered shelter to hibernating insects destroyed; no longer were crops rotated as frequently as good practice dictated. It was easier just to increase the dosage of insecticide in an attempt to counteract the aggravation of pest problems.

Rather than use biological control as an adjunct to chemical control, however, chemical control should be employed as an adjunct to biological and other types of control (to be discussed in the remaining chapters). Unfortunately, the consensus in entomological circles is that chemical control must come first. This attitude is explicit in all "official communiqués," as, for example, the brief of the Entomological Society of Canada and the official stance of the Entomological Society of America.

The greatest advance that could be made would be to remove pesiticides from the realm of commercialism. Yet, spokesmen in the United States Department of Agriculture say flatly: "There is no question at all that the primary responsibility for pesticide development and registration resides in the private sector." The private sector is industry. Pesticides are in the same category as patent medicines. Anybody can purchase them at the corner store or in bulk quantities from wholesalers.

Some thought should be given to the idea of regulating the sale of pesticides with the same stringency as applied to prescription drugs in the medical realm. There has been considerable resistance by U.S. governmental regulatory agencies to imposing restrictions, and the chemical industry is, of course, howling like a wounded bear. The Entomological Society of Canada has recommended that "The most dangerous pesticides should be sold on prescription and applied, preferably by licensed applicators, strictly according to instructions prepared by experts." This is a minimum recommendation. Optimally, *all* pesticides should fall under this interdict, especially since hazards are not always immediately apparent. DDT is an appalling example. The recommendation goes on to say that once a pesticide is registered there is little, or no, control over its use by individuals or governments, federal or local. It would be appropriate, continues the brief, to provide materials to homeowners only through outlets where they could also receive qualified advice, and to apply more through education and more selective licensing to applicators.

In some countries, governments are taking a more realistic attitude. Hungary, for example, exercises strict control over the *use* of agricultural insecticides. In Germany, a Plant Protection Law gives human and animal health priority over crop control of insecticides. The benefits of such legislation can extend beyond national boundaries because if a government stringently limits tolerable residues, growers that export to that country will be forced to find noninsecticidal means of control.

Generally speaking, the prevailing attitude is against restricting the use of insecticides. Whereas restrictions are being placed on the manufacturers, none are being placed on the users. Not only are private citizens conditioned to the idea that insects must be eliminated, they are barraged with propaganda whenever there is a population surge of any species. The question as to whether the insect involved poses any threat whatsoever is never honestly presented and more often than not can not be answered. The gypsy moth already referred to is of questionable danger, but once a population explosion is predicted, the news media, with grist provided by various industrial and governmental interests, begin to grind out their dire prophecies.

On March 21, 1971, the *New York Times* devoted nearly a full page to an article entitled "The Northeast Braces for Another Caterpillar Summer." Telling the people that New Jersey had sprayed 152,000 acres and New York 98,340 acres, and asking "Where does the individual property owner fit into the picture?," it left little doubt that towns and neighborhood groups should start some control and that spraying has first priority. Anybody from town officials down to the individual homeowner can douse as much Sevin as he can afford on the foliage.

A report on pesticides and ecology in the United Kingdom presented at the Annual Meeting of the Entomological Society of America in 1970 showed that users in the United Kingdom were still free to buy and apply insecticides as they chose. Information on use and hazards is available, but no control at all is exercised. The chaotic state of affairs is starkly evident in the statement that "non-agricultural uses may now be contributing a significant fraction [of insecticide] in the environment. There is a lack of information on the amounts of these chemicals used for non-agricultural purposes. . . ."

The consumer, whether he be the householder, the selectman of a town, or the farmer is constantly pressured by industry. A quotation from Paul DeBach is an accurate indictment of the system:

The average farmer has been thoroughly "sold" by insecticide salesmen, extension literature, and so-called economic entomologists. He has adopted the oft-repeated TV brainwashing slogan "the only good bug is a dead bug." Now, obviously, this has to change, not because biological control workers think it's bad, but because it doesn't work. Resistance to pesticides has developed, upsets have occurred time and again, and toxic residues have become a public health problem. The grower's psychology toward pest control is being forced to change, and this will force him to try other methods, which will mean greater support and use of biological control.

Meanwhile the consumer has aided and abetted the idea that "the only good bug is a dead bug" by thoroughly accepting the advertising idea that shiny clean fruit, etc. are better fruit. No thrips scars, not a scale insect, must be present. Quality and taste are really forgotten to a large extent; appearance is of prime importance. The Public Health Service also says that only a certain very small amount of insect parts can be bottled up, for instance, with your catsup. People aren't supposed to like small bits of insects in their catsup even if they can't find them. Thus, Public Health, which is trying to keep poisonous residues out of foodstuffs, may be forcing the tomato grower to treat, with the result that the consumer has a bug-free but not toxicant-free catsup.

Strict standardization procedures, which do not necessarily give the consumer food of better quality or taste, may force the grower to treat for a light infestation of a particular pest which would not be at all harmful to the crop. Thus, very strict standardization requirements may make the achievement of biological control a practical impossibility in particular instances. Even with the best examples of biological control, pest individuals are present. As food supplies become shorter during the coming decades and as consumers become more conscious of toxic residues in foods and less conscious of appearance only, certain artificial procedures, which now make insecticidal treatment a virtual necessity will be relaxed and more advantage will be taken of the biological control method.

The surface has only been scratched as far as the potential of biological control is concerned. There is still a realistic hope that insects for which parasites and predators have been sought for some time and which are still uncontrolled, such as the European corn borer and the gypsy moth, will capitulate if research is continued aggressively. Specialists in California do not believe that the point of diminishing returns is even in sight. The technique is gaining more favor, especially as failures in chemical control increase. More organizations are allocating funds to this area of research, but these continue to be niggardly in comparison with funds available for research and development of new insecticides.

In addition to the University of California, which continues to lead the field, the most outstanding organized efforts and studies are those of the Commonwealth Institute of Biological Control, which originated in England in 1927 and maintains its headquarters in Trinidad, British West Indies. While primarily a Commonwealth effort, it operates on an international scale. It has laboratories in California, India, Switzerland, Pakistan, and Trinidad. Other laboratories making major contributions are: the Canadian Department of Agriculture's Institute of Biological Control, with

headquarters in Ottawa and laboratories in Belleville and Van-
couver; the Insect Identification and Parasite Introduction Research Branch of the United States Department of Agriculture,
with laboratories in Moorestown, New Jersey, Beltsville, Maryland, and Paris, France; and the Hawaii State Board of Agriculture
and Forestry.

Recognizing that insects are apolitical beasts, a group of entomologists met in Stockholm in 1948 to discuss the need for an
international organization on biological control. As a result of
their deliberations, the Commission Internationale de Lutte
Biologique Contre les Ennemis des Cultures was established in
1949, and for the first time the cooperation of entomologists
interested in biological control was coordinated on an international
scale. The Commission is affiliated with the International Union
of Biological Sciences and is supported by contributions from
its member countries.

It has been remarked that there is no hope for biological and
other types of nonchemical control until the realms of agriculture
and economics become "uncoupled." A complete uncoupling is
perhaps unrealistic; however, our goals must change. As Corbet
pointed out at the International Congress in Canberra in 1972,
the chief constraint against integrated control (biological plus
other types) is the socioeconomic setting; namely, the idea of
maintaining high sustained yield with the least energy expenditure.
This idea is explicit in the program of the Food and Agriculture
Organization of the United States, whose aim is to increase the
yield of existing land and not bring new land into cultivation.

If the goal of agriculture does not change, the pressure on arable
land will increase, and impossible demands will be made upon the
plant-insect relationship. If the public's attitude toward insects
does not change, the same impossible demands will also be made.

References

1. Beirne, B. P., The biological control attempt against the European wheat stem sawfly, *Cephus pygmaeus* (Hymenoptera:
   Cephidae), in Ontario, Canad. Ent., 104 (1972), 987-90.
2. Beirne, B. P., Influences on the development and evolution
   of biological control in Canada. Bull. Ent. Soc. Canad., 5
   (1973), 85-89.
3. Brues, C. T., *Insect Dietary*. Harvard University Press, Cambridge, 1946, 466 pp.

4. Burges, H. D., and Hussey, N. W. (eds.), *Microbial Control of Insects and Mites*. Academic Press, New York, 1971, 861 pp.
5. Clark, L. R., Geier, P. W., Hughes, R. D., and Morris, R. F., *The Ecology of Insect Populations in Theory and Practice*. Methuen, London, 1967, 232 pp.
6. Clausen, C. P., Biological control of insect pests. Ann. Rev. Ent., 3 (1958), 291-310.
7. Corbet, P. S., Application, feasibility and propsects of integrated control. Address to 14th International Congress of Entomology. Canberra, Australia, August 28, 1972.
8. DeBach, P., (ed.), *Biological Control of Insect Pests and Weeds*. Chapman and Hall, London, 1964, 844 pp.
9. Doutt, R. L., Vice Virtue and the Vedalia. Bull. Ent. Soc. Amer., 4 (1958), 119-23.
10. Doutt., R. L., The historical development of biological control. In: *Biological Control of Insect Pests and Weeds* (P. DeBach, ed.), Chapman and Hall, London, 1964, pp. 21-42.
11. Doutt, R. L., and Smith, R. L., The pesticide syndrome-diagnosis and suggested prophylaxis. In: *Biological Control* (C. B. Huffaker, ed.). Plenum Press, New York-London, 1971, pp. 3-15.
12. Entomological Society of Canada, 1970. Pesticides and the Environment. Brief prepared at the request of the Board of Governors—1970. Suppl. Bull. Ent. Soc. Canad. 3 (1): 1-16.
13. Essig., E. O., *A History of Entomology*. MacMillan, New York, 1931, 1029 pp.
14. Faust, J. L., The Northeast braces for another caterpillar summer. The New York Times, March 21, 1971, D.35, D.38.
15. Faust, R. M., The *Bacillus thuringiensis* α-exotoxin: current status. Bull. Ent. Soc. Amer. (1973), 153-56.
16. Hansberry, R., Prospects for nonchemical insect control—an industrial view. Bull. Ent. Soc. Amer., 14 (1968), 229-35.
17. Howard, L. O., *A History of Applied Entomology*. Smithsonian Misc. Pub. Washington, 84 (1930), 1-564.
18. Huffaker, C. B. (ed.), *Biological Control*. Plenum Press, New York-London, 1971, 511 pp.
19. Klassen, W., and Schwartz, P. H., The role of the USDA in developing new chemical insecticides. Bull. Ent. Soc. Amer., 19 (1973), 98-99.

20. LeRoux, E. J., Biological control attempts on pome fruit (apple and pear) in North America, 1860-1970. Canad. Ent., 103 (1971), 963-74.
21. Metcalf, R. L., The impact of the development of organo-phosphorus insecticides upon basic and applied science. Bull. Ent. Soc. Amer., 5 (1959), 3-15.
22. Oliff, S., 1890. Agri. Gazette 1(2): 63-66.
23. Papworth, D. S., Pesticides and Ecology. Bull. Ent. Soc. Amer. (1971).
24. Rollins, R. Z., Drift of pesticides. Calif. Dept. Agric. Bull., 49 (1960), 34-39.
25. Simmonds, F. J., Biological control—past, present and future. J. Econ. Ent., 52 (1959), 1099-1102.
26. Simmonds, F. J., The economics of biological control. J. Roy. Soc. Arts, Oct. 1967, 880-98.
27. Steinhaus, E. A., Microbial control—the emergence of an idea. Hilgardia, 26 (1956), 107-60.
28. Steinhaus, E. A., Microbial diseases of insects. Ann. Rev. Microbiol., 11 (1957), 165-82.
29. Steinhaus, E. A., The importance of environmental factors in the insect-microbe ecosystem. Bact. Rev., 24 (1960), 365-73.
30. Steinhaus, E. A., (ed.), *Insect Pathology. An Advanced Treatise.* Vol. 1 and 2, Academic Press, New York, 1963, 1350 pp.
31. Taylor, T. H., Biological control of insect pests. Ann. Appl. Biol., 42 (1955), 190-96.
32. Turnbull, A. L., and Chant, D. A., The practice and theory of biological control of insects in Canada. Can. J. Sci., 39 (1961), 697-753.
33. Wade, N., Insect viruses: a new class of pesticides. Science, 181 (1973), 925-28.
34. Wilson, F., The Future of Biological Control. 7th Commonwealth Ent. Conf. Rpt. London, 1960, pp. 72-79.

Chapter Six

Birth Control, Reversed Eugenics, and Continence

6

Birth Control, Reversed Eugenics, and Continence

Man is trying desperately to limit two populations: insects' and his own. He strives to reduce the number of insects by increasing the death rate; he strives to reduce his own numbers by decreasing the birth rate. Chemical and biological control have been the chosen means for killing insects: chemical control for reducing populations to zero or very low levels temporarily; biological control for achieving low or tolerable levels held in permanent balance. Neither method has performed at its fullest potential; however, neither method is perfect even under the most ideal conditions, nor can either work alone. Chemical control has failed partly for lack of restraint; biological control partly for lack of money and effort.

The idea of applying birth control to insects is an old one, but not until recently has it been implemented. In theory, there are numerous procedures that can lead to a reduction in the birth rate. The most obvious one is to make insects sterile. Another way is to prevent them from reaching sexual maturity—the "Peter Pan" approach. A third is to prevent copulation. Each of these procedures can be varied. At first glance they may smack of science fiction; however, they can in fact be accomplished, and they do offer realistic ways of reducing the natural populations of insects.

The idea of employing sterility as a method for control first occurred to E. F. Knipling, an entomologist with the United States Department of Agriculture, in 1937. At that time, he and his colleagues were working in Texas with the screwworm. The screwworm, actually the larva of a fly (*Cochliomyia hominivorax*), is an obligatory parasite of warm-blooded animals in the Western Hemisphere, where it is a major pest of cattle. From its winter home in the American tropics it spreads each summer into Texas and several hundred miles northward. Until 1933, it was restricted to the country west of the Mississippi; but in that year it entered Georgia in a shipment of infested cattle from Texas. Soon it was in Florida. Since Florida is subtropical, the newcomers spent comfortable winters there. Each summer the flies multiplied in Florida and Mexico and spread northward throughout cattle country in an expanding wave of destruction.

The Department of Agriculture laboratory in Texas, seeking means of controlling this fly, had established laboratory breeding colonies. Two other entomologists, Melvin and Bushland, had succeeded in growing flies by the thousands on a nauseating mixture of meat, blood, and water. Observing the adult flies, Knipling noticed that there was a great deal of mating by three- and five-day-old flies but none by older flies. He guessed that females mated only once. It then occurred to him that it might be possible to control the screwworm fly in Florida if the native population could be overwhelmed with millions of males made sterile in the laboratory. By sheer numbers, these would mate preferentially with the females and no fertile eggs would be produced.

This was an ingenious and revolutionary idea, but many difficulties stood between the idea and its application. Were the females indeed monogamous? How could males be sterilized and released? Would they have normal sexual behavior and go about the business of procreation with gusto? To these question there were no known answers at the time.

In 1916, a report had been published to the effect that cigarette beetles produced infertile eggs after exposure to roentgen rays (X-rays), but neither Knipling nor his associates were aware of this research. In the meantime, Knipling and his associates were transferred to other work, and the idea lay fallow for nearly ten years. In 1950, another Department entomologist, A. W. Lindquist, called to Knipling's attention a recently published article by the geneticist Muller which mentioned that dominant lethal mutations resulting in sterility could be produced by X-rays. Knipling wrote to Muller outlining his idea for control and soliciting an opinion as to its feasibility. Muller had reservations but encouraged Knipling to conduct some experiments. These quickly revealed that either x-radiation or gamma radiation of pupae produced sterile flies that were in all other respects normal.

The next question to be settled was whether enough flies could be released to outnumber a native population. In 1952, on Sanibel

Island off the coast of Florida, large numbers of laboratory-bred flies were released. These flies had been raised on meat or cattle containing radioactive phosphorus (^{32}P). After the release, a survey of all flies on the island was made to ascertain how effectively those from the laboratory had infiltrated the population. From this survey, which showed which flies were radioactive and which were not, it was concluded that 100 sterile flies per square mile released each week could soon replace the native population. Accordingly, these numbers were released. In short order, the screwworms were eradicated. The island was not permanently freed, however, because of immigration by fertile flies from the mainland. This field test demonstrated, nevertheless, that Knipling's basic idea was sound. A second test on the island of Curaçao, in which 400 sterile males per square mile were released, resulted in the eradication of the screwworm from that area.

Up to this point, the tests had been restricted to relatively small areas; Curaçao embraces only 170 square miles. The next test was more ambitious. It involved an area of 2,000 square miles along the coast of Florida, in the vicinity of Cape Canaveral. This test, too, was successful. The stage was now set for a still more ambitious program to eradicate the screwworm in the entire state of Florida. By July 1958, facilities had been perfected for producing 14 million flies per week; its eventual capacity was 60 million per week. The Florida peninsula was divided into flight lanes twelve miles apart. Pilots flew six days a week. By staggering their rest days, it was possible for them to release flies every day.

Each day, new flight lanes were set so that, by the end of two weeks, the area under treatment had been covered by contiguous one-mile swaths. In this way, eight hundred sterile males and females were released per mile per week. About 50,000 square miles were treated at a cost of $10.6 million, resulting in an estimated savings in livestock of $20 million annually. So successful was the program judged to be that an area of comparable size was treated in the southwestern corner of the United States; a 1,500 mile-long barrier zone maintained between the United States and Mexico was also treated to intercept Mexican-born migrant flies crossing the border. The continuing cost is $6 million annually, but since 1966, there have been no resident screwworm flies in the United States.

The spectacular success of the screwworm program raised hopes throughout the world of economic entomology and stimulated a massive research effort. The technique seemed tailor-made for controlling the many kinds of tropical fruit flies that lay their eggs in oranges, apricots, cherries, peaches, and olives, resulting in fruits well-laced with maggots. Among the most destructive of these rather handsome flies are the Mediterranean fruit fly (*Ceratitis capitata*); the Mexican fruit fly (*Anastrepha ludens*); the Oriental fruit fly (*Dacus dorsalis*); the melon fly (*Dacus curcurbitae*); the olive fruit fly (*Dacus oleae*); and the cherry fruit fly (*Rhagoletis cerasi*).

Under certain conditions, the sterility principle is effective against them. In pilot tests with the melon and Oriental fruit flies in the Mariana Islands, eradication was achieved; however, the populations were low and geographically isolated on islands, and the control was integrated with other methods. Sterile males have also been used with some success in countering the migration of Mexican fruit flies into California.

Control of the Mediterranean fruit fly has been less successful so far partly because it is less tolerant of radiation and less sexually competitive after treatment. Nevertheless, intense effort has gone into perfecting the technique for this fly. In 1967, for example, a facility in Costa Rica was able to produce 40 million flies per week. The cost of rearing gradually fell from $50 million per million flies to $15 million per million flies. Improvements are constantly being made in techniques for rearing, treating, transporting, and releasing flies; however, more field tests are needed before the method can be accepted as standard.

In the meantime, best results are being obtained by combining the release of sterile males with other methods. In the Mariana Islands, for example, success was achieved on the island of Rota in 1962 against melon flies after the fly population was first partially reduced by distributing poisoned baits on farms. Flies were attracted to a protein hydrolysate containing insecticide. On Guam, advantage was taken of a reduction in the population of Oriental fruit flies caused by hurricanes. Then poisoned baits were distributed, and, lastly, lures that attracted and trapped males. Only then, when the population was at its lowest ebb, were the sterile flies released. In Saipan and Tinian, where no supplementary methods were employed, eradication was not achieved.

In Spain, pilot programs against the Mediterranean fruit fly reduced the infestation on the island of Tenerife from 70 percent to 15 percent and the infestation of citrus, apricot, and peach in the province of Murcia from 60 percent to less than 1 percent. In these last two instances, the areas treated were small. The programs did reveal, however, a rather special problem: Although females were sterile, they still went through the motions of laying eggs. As a consequence, many punctures were made on fruit, which permitted the entry of plant pathogens and which rendered the fruit cosmetically unacceptable on the market.

Other interesting successes have been achieved. The cockchafer (*Melolontha vulgaris*) is a beetle whose grubs live in the ground, feeding on grass; the adult beetles destroy other kinds of vegetation. The cockchafers live in isolated colonies as grubs, and, upon emergence, fly to the highest skyline on the horizon. A successful control technique used against them is that of digging grubs out of the ground in a place where their presence is of no importance, irradiating them and then releasing them in some area where eradication is desired (Horber). There the defective beetles mate with the local residents and the population is thus eliminated.

In confined situations, as in stored grain, complete infestation can be eliminated by irradiating entire populations—a much cleaner technique than treatment with insecticides.

T. Jermy, Director of Hungary's Research Institute for Plant Protection, has said:

When comparing the dramatic and continuing increase in published work on the sterile-release technique to the number of agricultural pests which are already controlled in practice by this method, one has to conclude that after the very encouraging initial practical successes achieved by American entomologists in the 1950s and early 1960s, the most recent achievements in practical control bear no satisfactory relation to the increasing efforts made in this work throughout the world. The causes of this slackening in the pace of development are more or less known to those engaged in sterile-release projects, but they are much less known to the agencies and authorities who decide on the distribution of research funds. The situation is made worse by the fact that in many cases the feasibility of the sterile-release technique has been overemphasized by authors of popular scientific works, and sometimes oversold by the scientists themselves.

It is the same old story.

The hopes raised by the success of the screwworm program have been tempered by the appearance of a number of unanticipated problems—some rather obvious, others subtle and complex. The scarring of fruit is only one. There is also the economic problem resulting from the expense of breeding and releasing huge numbers of insects and the sophisticated techniques required; however, it may be expected that costs will decline as techniques become more standardized.

There are also biological problems. In the first flush of success, it was hoped that the techniques could be applied to any pest that existed in a geographically isolated population or in very large areas where there would be no minimal migration of fertile males from adjacent untreated areas. However, insects, as usual, prove to be wily adversaries. They constitute such a large and evolutionarily diverse assemblage of species that there are no universal weaknesses. Each is practically a law unto itself. In order to play into the hands of the entomologist, a sterile male must be as vigorous and competitive as his fertile colleague. Many insects are so debilitated by sterilizing radiation that they make a poor showing on the field of honor.

Sterility by radiation arises from a dominant lethal mutation caused by chromosomes being broken. In flies, each chromosome is attached at only one point to the fibers that will draw it into the daughter cell at the time of cell division; therefore, if radiation breaks the chromosome, unattached fragments are lost. The resulting egg or sperm is defective. In moths, butterflies, and true bugs there are many attachments on the chromosome, so that even if the chromosome is fragmented, all pieces are drawn into daughter cells. Furthermore, the amount of radiation necessary to fragment these chromosomes is greater than that required with flies; hence, damage occurring in other tissues affects the metabolism or behavior of the individual. Sterilization would be a superb way to control the cotton boll weevil, for example, since by living inside cotton bolls it is immune to treatment with insecticides; however, irradiation so damages its gut that treated males are quite ill and noncompetitive.

Another difficulty arises from the differences in the kinds of effects that irradiation may have on reproductive tissue. Under one set of conditions, the sperm cells that are developing in the testes are inactivated, and the sperm-producing cells (spermatogonia) are killed. Accordingly, males so treated are truly sterile and produce no sperm cells. If these males are to be successful agents for control, they must compete in the field with normal males. Thus, the technique is useful only in controlling those species of insects that are monogamous, and then only provided that there are more sterile males than normal males. On the other hand, irradiation sometimes affects only the developing sperm, not harming the testis itself. The sperm cells are lethal, but they are not inactivated. The testis is later repopulated with normal sperm. The male is only temporarily sterile; however, during the tenure of his sterility, he is producing active but useless sperms. In this case, competition in the field is not between males but between sperms.

The best of all possible conditions would be that in which existing sperm cells are lethal but active, and the sperm-producing cells of the testis are killed so that no normal sperms will be produced subsequently. Under these circumstances, the sterile-male technique ought to be effective with polygamous as well as monogamous species of insects. The stumbling block posed by radiation damage tempered enthusiasm; however, hopes were revived when new techniques were reported.

In 1960, La Breque and his colleagues discovered that certain complex chemicals fed to insects induce sterility. Among these early components were aziridnyl derivatives of tetramine and morzid. Since 1960, 8,000 additional compounds have been screened for their effectiveness as "chemosterilants," and many have been found to be very powerful. They can be fed to the larvae or to adults. In females, some cause various kinds of abnormalities and degeneration of the ovaries and developing eggs. Others permit the development of eggs but interfere with the normal development of embryos. Still others permit full development of embryos, but the larvae do not survive after hatching. Even though chemosterilants are less effective in males, some do interfere with the development of sperm cells or their mobility, if they do develop.

Although chemosterilants abolish some of the problems posed by irradiation, such as sterilizing without reducing competitiveness, they do not reduce the high cost of breeding insects for release. One attempt to overcome this problem is illustrated by a pilot study on the feasibility of autochemosterilization of tsetse flies in Africa. These flies are readily attracted to moving objects, vehicles as well as large mammals. In the study, a bullock was fitted in an overall that had been impregnated with a chemo-sterilant. As the animal was driven through the bush, flies attracted to it were contaminated when they landed on the overall. It was not safe to spray the bullock directly because of the toxic and mutagenic properties of the chemosterilant (the cause of some of the drawbacks inherent in autochemosterilization). In addition to toxicity, many of the compounds have long persistence in the environment. In short, they may be the heirs to all the undesirable characteristics of insecticides.

Probably the most subtle and insidious drawback to the sterile-male technique is the effect that evolution itself will produce. No matter what man does to the environment and its inhabitants, he cannot escape the slow, inexorable march of evolution. It is not surprising, therefore, that predictions derived from studies of mathematical models indicated that natural selection would begin to counteract the effects of saturating a population with sterile males. It was predicted that so long as some normal males remained in the environment, there would be a natural selection favoring those females that did not interbreed with the defective males. It was further predicted that at the Texas-Mexican border, where there is no eradication of the screwworm fly but only a barrier zone separating Texas from normal breeding populations farther to the south, there would evolve Mexican flies that would ignore completely the laboratory-bred sterile males. Since those predictions were made, the Screwworm Eradication Program has indeed run into trouble. Some of the trouble has been traced to evolutionary changes, but as R. C. Bushland (3), one of the pioneers of this research, has said: "I regretfully agree—that no one has gotten to the bottom of the screwworm problem, but we are digging."

INSECT EUGENICS

The future is not, however, completely without promise. The fact that sterility, whether induced by irradiation or by chemical treatment, occurs because chromosomes are damaged suggests that genetic engineering might be used to control or suppress populations of pests. After all, we humans have talked a great deal about weeding out of our own population the genetically unfit, and the idea of improving mankind by practicing eugenics is not a new one. Why not, then, apply the reverse of eugenics to insects? Flood the population with undesirables and misfits. Create artificial species, species that would replace pest species and either become innocuous or gradually die off because of their unfitness. In other words, instead of trying to kill existing populations of pests, infiltrate their ranks with a few genetic undesirables and take advantage of nature's powerful mechanisms regulating courting, mating, and reproduction in order to saturate existing populations with bad genes. Let the insects be the agents of their own destruction. This calls for some very basic knowledge of the genetics of insects and some marvelously tricky bio-engineering. In fact, many programs of research are already in progress in this new field, and, at least in theory, the prospects appear bright.

The health, adaptive fitness, and competitive spirit of an organism depend on a full set of chromosomes with a corresponding total balance of genes. Males become sterile because radiation causes chromosomal imbalances. Dominant lethal mutations are induced. This kind of mutation can be contrived genetically in the laboratory; strains so affected can be normal in all respects except that they produce throughout their lives lethal eggs or sperms.

A promising alternative to complete sterilization is delayed sterilization. In 1940, a Russian geneticist, Serebrovskii, pointed out that a type of abnormal chromosome (translocation) could be used to reduce populations of insects. For thirty years, the report was unknown in the western world; it only came to light when American entomologists at an international meeting in Vienna proposed the same technique. Simply stated, the technique consists of producing genetic individuals that are partially sterile and introducing them into the field where partial sterility will be inherited from generation to generation. The population would tend to drift back to normalcy unless the genetic conditions for sterility were reintroduced from time to time.

As already mentioned, irradiation of moths, butterflies, true bugs, and some other kinds of insects, at dosages that do not render the insect totally unfit, breaks up the chromosomes as usual; however, because of the way in which fibers are attached to the chromosomes, the fragments are not lost. They are drawn into the daughter cells. Here they reassemble, but in such a way that the linear order of genes is pied. In technical terms, the eggs and sperms have heterozygous translocations. The fertilized eggs live and develop into first generation offspring that are sterile. For example, male moths exposed to less than 50 percent of the radiation required to sterilize them father progeny that are 90 percent sterile. Furthermore, these odd males are even more competitive than their normal colleagues, so they do an excellent job of saturating the population with their defective sperms.

Even more bizarre results can be obtained with other models of genetic engineering. One is called sex-ratio distortion. Normally the two sexes are fairly evenly distributed in a population. If the proportion of one sex could be drastically reduced, as happened in the human population in some parts of Europe after World War I and again after World War II, the rate of population growth could be impeded to the point where complementary control measures could become effective. This has actually been accomplished in the laboratory with houseflies. A strain has been developed in Australia the males of which, when mated with houseflies from other parts of the world, produce offspring that are 90 percent males. Another technique takes advantage of the so-called frigidity factor. When copulation occurs in such a monogamous species as the housefly, the seminal fluid of the male causes the female to repel all subsequent attempts at copulation. This works even if the males have been castrated. For control, all that would be necessary would be to release competitive sterile males (these males would have no viable sperms, but would still have seminal fluid).

Still another kind of genetic freak is the conditional lethal mutation. An animal containing such a mutation is normal, fit, and competitive under normal environmental conditions, but fails to cope when conditions are altered. For example, an inability to hibernate (diapause) could be a grave handicap when seasonal conditions for normal living become impossible. Similarly, inserting into the population a mutant sensitive to high or low temperatures would decimate the population when temperature extremes occurred. The trick is to choose a heritable characteristic that becomes lethal when some normal environmental stress such as desiccation, high humidity, heat, cold, or length of day (photoperiod) becomes restrictive.

Mutations of this kind have been referred to as "genetic time bombs." These mutations could also be combined with mechanisms for reducing the genes for resistance in a population that no longer responds to insecticides. This can be accomplished genetically in a few generations without having to release billions of flies. The sex booby-trap is another alternative. This technique involves breeding and releasing sterile females that are highly resistant to insecticides. Each one is thickly coated with an insecticide before it is released. Every male that courts it is lethally poisoned.

Still other kinds of mutations can be produced. Insects that for genetic reasons will not eat their normal food can be introduced into a population. These, passing on their genes to the natives, will eventually cause the development of a self-starving population. Even more imaginatively, a population of pests could be genetically engineered so that it would eat weeds instead of crops. In theory it is all possible; in practice there are many obstacles, and the artificial "species" produced by geneticists will always be up against the normal forces in evolution that may be the instruments of quite other designs. The limiting factor, the bottleneck in the exploitation of genetic engineering, is, according to the experts, the scarcity of competent geneticists interested in studying these methods of pest control.

Concurrent with the efforts of geneticists to contribute new approaches to the control of insects are those of endocrinologists. Ponce de León may have searched unsuccessfully for the fountain of youth, but endocrinologists have found it—for their insects. In theory, the discovery offers a bizarre way of controlling the populations of undesirable species. A perpetually young, preadolescent insect can never reproduce.

A method of achieving perpetual youth evolved from a long series of studies that were motivated solely by a search for knowledge for its own sake—the purest of pure research. It stemmed from a curiosity about the almost miraculous transformation of the unglamorous, earthbound caterpillar to the transcendently beautiful butterfly—a kind of transformation that characterizes insects belonging to many orders, chief among which are moths and butterflies, beetles, flies, ants, bees, and wasps. The first significant step in this train of events was the discovery in 1920 by the Polish biologist Stefan Kopěc that there was a hormone produced in the brain of the caterpillar (he worked with the gypsy moth) that controlled the change to a moth. Kopěc tied a thin ligature around the body of a caterpillar just behind the head in such a way that the caterpillar remained alive. When the front end of the beast changed to a pupa (the stage intermediate between caterpillar and moth), the hind end remained a caterpillar. Kopěc reasoned that some blood-borne substance (a hormone) was released from the anterior end. Later, by surgically removing brains and replanting them, he was able to pinpoint the source as existing in the brain.

In 1934, the English insect physiologist V. B. Wigglesworth, working with a blood-sucking bug, *Rhodnius*, discovered a second hormone present in immature insects but absent when the insect molted into pupa or adult. It came from a pair of small glands situated more or less close to the brain, the corpora allata. This was called the "juvenile hormone." In 1941, a third hormone originating in the anterior part of caterpillars (silkworms) was discovered by the Japanese Fukuda. This hormone, whose action prevented molting and pupation, was called by Fukuda *ecdysone* (molting hormone, from the Greek *ekdysis*, an escape). It originated in the prothoracic glands, a pair of vermiform glands in the first segment of the body. Meanwhile, an American, C. M. Williams, working with a giant American silkworm, the *Cecropia*, demonstrated that another hormone came from 26 neurosecretory cells in the brain and that it triggered the production of ecdysone from the prothoracic glands. Ecdysone initiated molting.

The sequence of hormonal events leading to normal growth and development in insects can be summarized as follows. At the beginning of life, at the time of hatching from the egg, brain hormone triggers the release of ecdysone. At the same time, juvenile hormone is released in a controlled fashion by the corpora allata. For normal growth and successive molting of immature insects, both ecdysone and juvenile hormone must be present. When the immature insect reaches the end of its larval or nymphal development, the flow of juvenile hormone decreases, and, in its absence, ecdysone causes the insect to pupate and eventually to emerge as an adult. Ecdysone is necessary for all kinds of molting; it stimulates the necessary biosynthetic activity. Juvenile hormone regulates the *kinds* of biosynthetic activities that occur in response to ecdysone.

When insects are injected with ecdysone, development is accelerated, and they die prematurely old; if they are injected with juvenile hormone, they remain perpetually juvenile or, at best, develop into some forms intermediate between juvenile and adult. In vertebrates, sexual maturity occurs in response to the secretion of maturation hormones, and sexual immaturity exists in the absence of these. In insects, by contrast, the juvenile state is maintained by the juvenile hormone, and sexual maturity is brought about by its absence. In short, these hormones have the potential for interfering with development in such a way that populations can be controlled.

The task of identifying and synthesizing the hormones is formidable because they are never present in the insect body in large amounts. They are, in fact, produced only as needed and not stored in advance. It required 2,000 pounds (1,000 kilograms) of silkworm pupae to produce a tiny sample (250 milligrams) of the first pure ecdysone. The task was accomplished in the 1950s by the German biochemists Adolf Butenandt and Peter Karlson. Not until ten years later was the structure identified and a sample synthesized. This was accomplished simultaneously and independently by two teams of industrial chemists, one at Syntex Research in California and the other at Scherring and Hoffman-La Roche. The Japanese, headed by H. Mori, also synthesized ecdysone. Schneiderman, Meyer, and Gilbert spent seven years and 40,000 bioassays isolating juvenile hormone from half a million Cynthia moths raised in Japan and 70,000 Cecropia moths in the United States; but the final identification of Cecropia juvenile hormone was accomplished in 1965-66 by Herbert Röller at the University of Wisconsin. Previous to that, analogues had been isolated or synthesized by Schmialek in Berlin, and Law and Bowers in America. Three other American groups synthesized juvenile hormone in 1968 and in 1969 additional syntheses were accomplished in Japan, Germany, Canada, and Australia.

The next link in the chain of events was provided by Karl Sláma of Prague. Linden bugs (*Pyrrhocoris apterus*) that he brought to America to study refused to mature; they became giant adolescents. By meticulous detective work, Sláma found that American paper used to line the bugs' cages contained something that mimicked juvenile hormone. When the bugs were raised in the absence of American paper or in cages lined with European paper, development was normal. Further detective work revealed that the chief source of wood pulp in Canada and New England is the balsam fir (*Abies balsamea*). In 1966, William Bowers in the United States

Department of Agriculture's laboratory at Beltsville, Maryland, isolated and identified the elusive factor, showed that it was closely related to juvenile hormone, and named it *juvabione*. Curiously, it is effective only against European bugs related to *Pyrrhocoris*. American insects are immune.

The importance of the "paper-factor saga" was that it suggested that plants might produce compounds mimicking insect hormones. Sláma's group, and another group at Tohoku University in Japan, soon showed that the plant world is indeed rich in mimics. Evergreens and ferns in particular are rich in ecdysone or its mimics. Tests have been run on more than 1,000 species of plants. Forty have shown hormone activity, and more than 25 phytoecdysones have been discovered. All have structural similarities; they are steroids. By contrast, the many analogues (composites structurally similar to the hormone) and mimics of juvenile hormone that have been and are still being discovered in diverse animals, microorganisms, and plants represent a wide variety of unrelated compounds.

Because hormones and their mimics seriously interfere with development, the strategy for control of pests is to alter the rate of growth so that insects develop at the wrong time of the year, or to interfere with egg development, embryonic development, metamorphosis, or molting. For example, since many insects spend the winter in a state of arrested development, the application of brain hormone, which controls this pause in development, would cause insects to forego hibernation and continue their activities until winter and lack of food overtook them.

Ecdysones offer more feasible approaches to control because they affect insects at all stages of life. They cause precocious molting; some in certain dosages are chemosterilants; some are toxic. They apparently are non-toxic to vertebrates. Furthermore, since they are of widespread occurrence in nature anyway, it is difficult to imagine how they can have major adverse environmental effects. Much more work remains to be done, however, before one could safely disseminate these compounds.

Juvenile hormones may be even more effective as agents for control. They too appear to be non-toxic to vertebrates and benign insofar as the environment is concerned. The potency of chemicals similar to juvenile hormone is incredible. Sláma and his colleagues have shown that 0.0001 gram will cause a *Pyrrhocoris* larva to change into a form that resembles an adult but is unable to reproduce and survive. This is equivalent to 2.5 milligrams for 25,000,000 insects. Since a dose of only 0.0001 (1μ) per square meter of filter paper is effective, 10 milligrams per hectare would give effective control. Schneiderman pointed out, however, that a hectare of soil is very different from a hectare of filter paper, and practical problems of field application have not yet been solved. He added: "Another problem which should not be ignored is the fact that many of the potent juvenile hormone analogues are stable chlorinated hydrocarbons which may have ecological side effects similar to those of chlorinated insecticides like DDT. However, the results up to this point are promising and seem to warrant intensive study by applied entomologists." Control by hormones has stimulated the interest of industry, and a company in California, Zoecon, was formed specifically for the purpose of developing hormonal materials for control.

Thus, although hormones, their analogues, and their mimics offer great promise, they are still an unknown quantity. They appear to be non-toxic to vertebrates and non-detrimental to the environment. Some are extremely specific in their action and would not harm beneficial insects. Juvabione, for example, affects only *Pyrrhocoris* and its relatives. It was at first thought that there would be no problem with resistance because insects would not become resistant to their own hormones, but this dream has been shattered. Insects normally have mechanisms to inactivate their own hormones at specific periods during their development, and natural selection probably could favor strains that could inactivate externally-applied hormone-like compounds.

There are other problems. To be effective, juvenile analogues must act at certain critical times in the life of an insect. Since not all individuals in a population grow at the same rate, the compound must be stable enough to persist in the environment for several weeks if is is to affect all individuals. Unfortunately, many of the compounds are rapidly inactivated by moisture and by light. Furthermore, many of these compounds do not pass readily into the insect either through the cuticle or through the gut. Their synthesis, too, is still a difficult chemical problem, and yields are disappointingly low.

The present status for this promising form of control has been succinctly summarized by Ellis, Morgan and Woodbridge in the following words:

In spite of the more hopeful turn of events, there are still formidable problems to be overcome before any of the mimics can be used as a pesticide. The problem of persistence remains unanswered. Their possible adverse effects on non-pest species still cannot be evaluated from the present evidence. With the lesson of the organochlorine pesticides before us, very thorough investigations on safety would no doubt be demanded, and rightly so, before the hormone mimics could be used for pest control. Related to this is a lack of adequate studies on the degradation of these hormones and their mimics within the body. If more were known about this problem, it might be easier to assess the possibility of the development of pest species with strains that are resistant to hormone treatment. However, biologists will certainly continue their research in this field, and many of the answers will come from the results of pure research carried out in the universities.

CONTINENCE

The final approach to the task of insect birth control that we shall discuss in this chapter is continence. Insects can be tricked into seeking false sexual partners. This is made possible by the fact that the two sexes in most insects find their trysting places and conduct their courtship under the direction and influence of enticing scents, perfumes, and aphrodisiacs. There is no more graphic picture of the action of sex attractants, of chemical communication, than that given by Jean Henri Fabre describing the assembling of males of the Great Peacock moth, Europe's largest, to a captive virgin female:

We enter the room, candle in hand. What we see is unforgettable. With a soft flick-flack, the great moths fly around the bell-jar, alight, set off again, come back, fly up to the ceiling and down. They rush at the candle, putting it out with one stroke of their wings; they descend on our shoulders, clinging to our clothes, grazing our faces. The scene suggests a wizard's cave, with its whirl of bats.

Coming from every direction and apprised I know not how, here are forty lovers eager to pay their respects to the marriageable bride born that morning amid the mysteries of my study.

We ourselves, creatures of vision, can appreciate only dimly the importance of chemicals that direct the comings and goings of insects, which are creatures of scents and tastes, endowed with receptors of exquisite sensitivity. It is no wonder that Fabre and two generations of naturalists who followed him were at a loss to explain insect assembling.

Faulty observation and poorly designed experiments led to an exaggeration of the distances over which males assembled and to the impression that males could orient downwind of the female as well as upwind. Such beliefs led, in turn, to hypotheses involving emanations, infrared, radio waves, and other electromagnetic phenomena. Infrared theories were resurrected in the mid-1950s when they were given wide publicity in the press. More recently, electromagnetic radiation has again become fashionable in some quarters. Despite these cyclic perversions, it is clear that specific chemicals or mixtures of chemicals guide the male to the female and in some cases the female to the male.

Let us digress briefly at this point to clarify some terms that are encountered in the literature. Thirty years ago, compounds that influenced the behavior of insects were divided simply into two classes: attractants and repellents. Whenever insects were observed to congregate around a substance, that substance was called an attractant. As we learned more about the behavior of insects, it became apparent that their actions with respect to chemicals are too complex to be understood in simple terms of attractants and repellents only. We now speak of attractants, arrestants, feeding stimulants, mating stimulants, ovipositional stimulants, deterrents, and repellents.

An attractant is a substance that causes an insect to walk or fly to it in an oriented way from some distance, however, short. The scents of flowers and the products of fermentation in beer and spoiled fruits are attractants. An arrestant is a substance that causes insects to halt when they come into contact with it, irrespective of how they got there. If more than one insect is involved, aggregations form. Sugar is an arrestant par excellence. Flies, for example, are not attracted to it, but once they have stumbled upon it they remain. Stimulants are those substances that cause insects to begin eating, copulating, or laying eggs. Deterrents are those materials that prevent insects from eating or laying eggs, without necessarily driving them away. Salt on a cabbage leaf would prevent a cabbage caterpillar from eating; accordingly, it is properly a deterrent. A repellent causes insects to move actively away from the source. Most mosquito repellents exemplify this class of substances.

To return now to attractants—chemical sex attractants, first observed acting on virgin giant silkmoths and related species, are now known to occur in no less than forty-six species of insects, among which are cockroaches, true bugs, lace wing flies, scorpion flies, moths and butterflies, sawflies, ants, bees, wasps, and beetles. The phenomenon may be even more widespread than we realize because there must be some distance orientation mechanism in all insects to bring male and female close enough to see or touch one another. Imagine a boll weevil in one corner of a cotton field trying to find one in another corner without any guidance.

Proximity does not necessarily breed familiarity. Once the sexes have met, receptivity and the level of sexual excitement required for successful copulation may occur only under the influence of aphrodisiacs. Some extraordinary mechanisms have evolved. For example, the male Grayling butterfly (*Hipparchia semele* L.) during courtship clasps the antennae of the female between his wings, where they are stimulated by special scent scales. Male *Danaus* butterflies (e.g., the Monarch) evert hair-pencils on the end of the abdomen and dip them into scent pockets on the wings several times a day to steep them in scents that sexually excite the female. Many male moths also possess abdominal scent brushes which are everted as the males approach the female in response to her sex attractant. Enveloped in a cloud of male scent, she capitulates.

Modern chemical techniques have made possible the extraction and identification of many sex attractants and aphrodisiacs. They are now often referred to as *sex pheromones* (from the Greek "pherein," to carry, and "hormon," to excite), compounds secreted to the outside by an individual and received by a second individual of the same species, which then responds in a specific manner. As of 1965, one hundred and fifty-nine female sex pheromones were known and fifty-three male sex pheromones.

The pure compounds, singly or in carefully formulated mixtures, are, like the virgin females (or males) themselves, extraordinarily powerful attractants. A few molecules can attract and excite. Why not, then compete with the sexes in nature by luring swains into traps or by so completely saturating the air with scent that accurate orientation to the opposite sex is impossible? Success with either method would bring about a reduction in offspring. The idea seems elegantly simple, but matured slowly. In any case, the earliest application of attractants to agricultural problems relied not upon sex attractants but upon feeding attractants. Many valuable lessons were learned in the process.

The earliest recorded use of attractants for control was the suggestion of Pliny that a fish be hung in a tree to lure ants away from the foliage. No further attempts to make use of attractants were recorded until the eighteenth century. The first attempt by a professional entomologist was Coquillett's incorporation of attractants into baits to control grasshoppers in 1885. Nineteenth-century wine growers in Europe tried erecting traps baited with stale beer, old cider, or lees to control grapevine moths. By 1896, the attracting value of baits of rum, stale beer, or brown sugar was widely recognized by insect collectors. In the 1920s, hundreds of aromatic compounds were tested in a search for something more attractive to orchard pests than their natural food. In 1927, additional impetus was given to the search by the discovery of the extraordinary attraction that geraniol exerts for Japanese beetles.

Many powerful lures were discovered, but the hope that insects could be attracted away from food plants in sufficient numbers to save the crops or reduce the population was not realized. Trapping with lures was an effective means of sampling populations and of detecting new infestations before they became full-blown, but it did not work as a control measure. Feeding attractants simply could not compete successfully with food, they could not cause a significant reduction in the population, and they had the unhappy tendency, when they were very potent, of drawing insects from adjacent areas.

Although attempts were made as early as 1893 to control gypsy moths by employing the scent of females, intensive studies of sex scents as agents for control did not begin until 1913. Between 1913 and 1931, crude extracts of female abdomens were placed in traps employed in surveying new infestations. Around 1925, attempts to isolate attractants were begun at Harvard University. Soon the work shifted to the United States Department of Agriculture, where Collins and Haller and their colleagues attempted unsuccessfully to isolate and characterize the attractant. During this period, workers in several laboratories randomly testing chemicals from the shelf occasionally discovered some that only attracted males of one insect or another.

As early as 1912, Howlett had reported to the London Entomological Society that oil of citronella attracted only males of some species of fruit flies and, later, that iso-eugenol and methyl eugenol each attracted males of different species. In the Union of South Africa, Ripley and Hepburn discovered chemicals that attracted only males of the Natal fruit fly. The three scientists proposed that the attractive chemicals mimicked female sex attractants. The real attractants eluded identification.

During the 1930s, in Germany, Butenandt was busy trying to extract the female sex attractant of the commercial silkworm moth from 7,000 female abdominal tips. He managed to collect 100 milligrams of impure attractant. After many attempts, Butenandt and Hecker finally succeeded in 1959 in isolating the pure compound from 500,000 virgin female abdominal tips and synthesizing it in 1961. They called it bombykol.

Meanwhile, attempts to isolate the female sex attractant of the gypsy moth continued in America and Europe. At last, after working thirty years, chemists of the United States Department of Agriculture announced the isolation, characterization, and synthesis of the attractant in 1960. They called it gyplure. It proved not to be the attractant. In 1969, the real attractant was identified and synthesized by Bierl, Beroza, and Collier. It was given the name disparlure.

With the identification of bombykol and the development of analytical techniques permitting chemists to work with minute samples, emphasis shifted further away from feeding attractants to sex attractants. Whether they alone can successfully reduce populations of pests is questionable; the final verdict from the field has not been rendered.

In 1930-31, a Czechoslovakian entomologist, Dyk, started field experiments aimed at controlling the nun moth (*Porthetria monacha*), a close relative of the gypsy moth, by fastening captive virgin females to tree trunks and surrounding them with sticky paper. For the next ten years various modifications of this technique were tested. In one test, 384,448 males were caught in forty-nine days in 480 traps set in 756.57 hectares.

Another test of the potentialities of sex attractants for control was conducted in California. Here synthetic pheromones attractive to the western pine beetle were placed in traps distributed in an area of twenty-five square miles. They captured approximately 730,000 beetles. It had been predicted on the basis of extensive population analyses and aerial photographic surveys that 750,000 beetles would develop that year. From these figures it would appear that the traps had captured 97 percent of the population.

Generally speaking, pheromones alone cannot suppress large populations. In parts of Texas, traps containing live males of the cotton boll weevil disseminating male sex pheromone captured a large enough number of overwintering weevils to suppress the population. In late summer, however, migration from outside swamped the pheromones. A similar result was obtained in a field test in South Carolina. If, however, attractants are mixed with poisons or chemosterilants, their value as agents for control can be enormous. By attracting insects to specific places, the attractant obviates the necessity of massive broadcasting of insecticides.

Feeding attractants, though not so behaviorally potent as sex attractants, also serve well in this capacity. Eradication of the Oriental fruit fly from Rota Island has already been mentioned. The procedure there was to drop from aircraft small cane-fiber squares containing the attractant methyl eugenol and a quick-acting insecticide. Even though the lure attracted only males, eradication was achieved by dropping squares every two weeks over a period of several months. In the first eradication of the Mediterranean fruit fly from Florida, a similar tactic employed protein hydrolyzates laced with malathion.

Apart from direct control, attractants, particularly sex attractants, whether disseminated by Judas individuals, purified extracts, or synthetic compounds, have served well indirectly. As early as 1913, they were used for detecting and surveying new infestations of various insects. In Czechoslovakia, sex attractant traps have been employed for surveying since the early 1930s.

In the United States, traps baited with crude extracts of female abdomens were in use for twenty years to detect gypsy moth infestations. Each year it cost about $25,000 to bait 60,000 survey traps with extracts from tips procured in Spain, Yugoslavia, and French Morocco. When disparlure was finally synthesized, the first lot, 30 grams, was enough to last for 50,000 years at the rate at which crude extracts had been used; the eventual cost, thirty cents per gram. Today, survey traps baited with twenty *micro*grams of disparlure keep constant watch on this insect, which thus far has been contained in a small corner of its potential range.

Today, traps with attractants are also deployed around various ports of entry in the United States to intercept and give early warning about pests entering the country. Between 1958 and 1964, an estimated $9 million for potential eradication were saved by this early warning system.

The tactic of luring one sex or another into traps or to poisons operates on the principle of the decoy—the signal is loud and clear, but death, not consummation, is the final result. Another strategy that has been proposed is that of making the signal meaningless by releasing so much pheromone that the scent of the live insect is overwhelmed and no individual can find its mate in the surfeit of attractant. This approach has actually undergone experimentation with gypsy moths. It does not appear too hopeful at the moment. Still another approach is to jam the signal by masking. Certain compounds, variously termed "antipheromones" and "anti-attractants," that mask sex attractants, have been extracted from bark beetles, among which they presumably shut off attraction when a sufficiently great number of individuals have been attracted.

If the application of attractants to the control of agricultural pests can be perfected, it offers appealing features. Attractants act only on a specific insect. Because of their specific action and their effectiveness at very low concentrations, it is not necessary to spray vast quantities onto the environment. Attractants and all other approaches to insect birth control offer great promise as supplementary methods for protecting crops.

References

1. Beroza, M. (ed.), *Chemicals Controlling Insect Behavior*. Academic Press, New York and London, 1970, 170 pp.
2. Beroza, M., and Knipling, E. F., Gypsy moth control with the sex attractant pheromone. Science, 177 (1972), 19-27.
3. Bushland, R. C., Screwworm eradication program. Science, 184 (1974), 1010-11.
4. Dethier, V. G., *Chemical Insect Attractants and Repellents*. Blakiston, Philadelphia, 1947, 289 pp.

5. Dethier, V. G., Barton, Browne L., and Smith, C. N., The designation of chemicals in terms of the responses they elicit from insects. J. Econ. Ent., 53 (1960), 134-36.

6. Ellis, P. E., Morgan, E. D., and Woodbridge, A. P., Is there new hope for hormone mimics as pesticides? Pans (Tropical Pesticides Research Headquarters and Information Unit, London) 16(3): 434-46, 1970.

7. Fabre, J. H., *The Life of the Caterpillar*. Dodd Mead Co., N. Y., 1916, 376 pp.

8. Force, D. C., Genetics in the colonization of natural enemies for biological control. Ann. Ent. Soc. Amer., 60 (1967), 722-29.

9. Foster, G. G., Whitten, M. J., Prout, T., and Gill, R., Chromosome rearrangements for the control of insect pests. Science, 176 (1972), 875-80.

10. Horber, E., Eradication of white grub (*Melolontha vulgaris* F.) by the sterile male technique. Radiation and radioisotopes applied to insects of agricultural importance. Internat. Atomic Energy Agency, Vienna, 1963, 313-32.

11. Jacobson, M., *Insect Sex Attractants*. Interscience Pub., New York, 1965, 154 pp.

12. Jermy, T., Genetic control experiments on the bean weevil, *Acanthocelides obtectus*. Radiation and radioisotopes applied to insects of agricultural importance. Internat. Atomic Energy Agency, Vienna, 1971, 349-54.

13. Johnston, J. W., Moulton, D. G., and Turk, A., *Advances in Chemoreception*. Vol. 1, *Communication by Chemical Signals*. Appleton-Century-Crofts, New York, 1970, 412 pp.

14. Knipling, E. F., Sterile technique—principles involved, current applications, limitations and future applications. In: *Genetics of Insect Vectors of Disease* (J. Wright & R. Pal, eds.), Elsevier, London, 1967.

15. Roach, S. H., Taft, H. M., Ray, L., and Hopkins, A. R., Population dynamics of the boll weevil in an isolated cotton field in South Carolina. Ann. Ent. Soc. Amer., 64 (1971), 394-99.

16. Schneiderman, H. A., The strategy of controlling insect pests with growth regulators. Mitteilungen Schweiz. Entom. Gesellschaft, 44 (1971), 141-49.

17. Schneiderman, H. A., Krishnakumaran, A., Bryant, P. J., and Schnal, F., Endocrinological and genetic strategies in insect control. N. Y. Agri. Exp. Sta., Geneva, N.Y. Proc. Symp. on Potentials in Crop Protection, 1969, pp. 14-25.

18. Serebrovskii, A. S., On the possibility of a new method for the control of insect pests. Zool. Zh., 19 (1940), 618-30.

19. Smith, R. H., and Von Borstel, R. C., Genetic control of insect populations. Science, 178 (1972), 1164-74.

20. Whitten, M. J., Genetics of pests in their management. In: *Concepts of Pest Management* (R. L. Rabb and F. E. Guthrie, eds.). North Carolina State Univ. Press, Raleigh, 1970, pp. 119-37.

21. Zoecon Corpn., Ann. Rpt. 1970, Palo Alto, Calif. 30 pp.

Chapter Seven
Living Menus

7
Living Menus

In their relation to insects and to man, plants are usually thought of as food; or, occasionally, with respect to man, as raw materials for construction and manufactured goods. Seldom are they appreciated—except, perhaps, by some botanists—as responsive, struggling, living organisms. Yet, if we are to understand the insect/plant relationship and to manipulate it in our favor, we must adopt the view that plants are organisms competing for life with other organisms—with plants, men, cows, and aphids—and that they are as responsive to biotic and abiotic pressures as are animals.

In this view, the successful individuals and species of plants are those that survive the inimical features of their environment long enough to produce others of their kind. The pressures are ever present and varied because the earth does not provide an infinite supply of the necessities of life. Each green plant must compete for a place in the sun, a substratum into which it can anchor against the buffeting of weather, air, and water, and which provides mineral nutrients for its roots. It must evolve efficient mechanisms whereby it can survive under different soil (edaphic) conditions. Different species adapt to different conditions: cold, heat, long days, short days, sun, shade, wind, calm, acid soil, alkaline soil, sand, clay, salt soil, fresh soil, land, water. Once adapted to a set of abiotic conditions, each individual must compete with other species that have similar requirements and with other individuals of its own kind. A most instructive lesson in the severity of this competition is gleaned by sowing a packet of seeds—radish, for example—and without ever thinning the bed, watching the seedlings grow.

Superimposed upon these pressures are those applied by parasites and predators, the multitudes of microorganisms, plants like the mistletoe, the dodder, and the strangler fig, and the invertebrate and vertebrate grazers. (It is extremely difficult to apportion the source of burden imposed by these divers agents partly because they interact in so complex a manner.) A tree suffering from a viral or fungal disease may not survive the same degree of defoliation by caterpillars that a healthy tree will; a tree attacked by bark beetles is immeasurably more susceptible to secondary infection by microorganisms than is a tree that is spared attack. A combination of drought and attack by locusts may be more detrimental than either alone; plants growing where soil conditions are marginal survive grazing by vertebrates less successfully than do vigorous plants growing under ideal conditions. It is a mark of their resilience—their sensitivity, responsiveness, and adaptability—that green plants survive at all.

In the course of evolution, plants have evolved bizarre and wondrous mechanisms for defense against their parasites and predators. Their avenues of escape are, not unexpectedly, quite different from those most commonly adopted by animals. Evasive action is constitutionally impossible; a plant cannot pick up its roots and run, and only in science fiction does it fight back physically. Emigration to safer habitats by successive generations by way of seed dispersal or other methods of propagation is, in the final analysis, an ineffective measure against highly mobile, climatologically adaptable predators. A plant's best defensive strategy is to become unacceptable to its enemies. At this point it is well to remember that, although a plant may be just food to a cow or sheep, it is more than that to insects: It may be food, cradle, home, and castle. A plant provides a place to live and a unique microclimate.

PLANT DEFENSES

To render itself unacceptable, a plant may adopt two kinds of strategic defenses: the one structural; the other chemical. While structural defenses are by no means unimportant, they are more limited than chemical ones. Some examples include tough cuticles, abundant pubescence, unmanageable leaf shapes. Some families of plants possess sharp bundles of needle-like crystals of calcium oxalate that are as troublesome to a leaf-eating insect as bones in fish are to us. That these crystals are indeed a deterrent to feeding can be shown by dissolving them in dilute hydrochloric acid. Leaves so treated become acceptable food. Another interesting example of physical protection is that afforded by the sharp teeth on the leaves of holly. Some caterpillars that refuse holly leaves as food readily accept them when the spiny edges are shaved smooth. There is a variety of rice that is protected from damage because of the physical action of its high silicon content. Rice borers can not make sufficient headway in the stems, and they wear out their mandibles trying. An unusual case of defense is that by which some leaf-mining larvae are rendered incapable of damaging snapdragons because they are crushed by the rapidly proliferating wound tissue.

No matter how we look at plants, they remain primarily a source of food—a green, living menu. A plant, in contrast to many other foods, is alive all the time it is being eaten by herbivores (man excepted). Its best protection is to become unpalatable. This end is achieved by the elaboration of chemicals which are, for one reason or another, unacceptable to a predator. The chemical armament must, however, suffice to repel all attacks, not alone those of insects, but also those of viruses, fungi, round worms, and herbivorous vertebrates. Additionally, chemical defenses may be employed by a plant to suppress the growth of other plants that would encroach upon its living space.

Plants have evolved a truly spectacular array of unusual and unique chemicals. Formerly, these were for the most part comprehended under the term "essential oils." They include thousands of compounds: alkaloids, glycosides, terpenes, tannins, alcohols, esters, acids, saponins, steroids, carotenoids, to mention only a few categories. These compounds are responsible for the characteristic flavors of different fruits and vegetables, the delicate and heady scents of flowers, the kaleidoscope of colors in the flower garden; they are essential components of the aromatic oils of clove and peppermint; the caffein of coffee; the medicinals such as quinine, cocaine, opium; the latex that forms rubber; and indigo dye, inks, tannins, resins, turpentine, etc., etc. The list is interminable.

In contrast to primary chemicals, proteins, carbohydrates, fats, nucleic acids, and others that are generally characteristic of all organisms and are directly involved in essential metabolism, the odd plant chemicals are secondary in that they are not, so far as we know, concerned with ordinary metabolism. Their distribution in the plant kingdom is sporadic.

The plant expends considerable energy synthesizing these

compounds. Since there is apparently no metabolic use for them, and since no organism can afford the luxury of energy expenditure that does not serve some purpose related to survival, much mystery has surrounded the odd substances. As early as 1888, a botanist, Stahl, suggested that secondary plant substances served to defend plants against predators and parasites. In the past few decades, a great deal of evidence in support of this hypothesis has been accumulated. The subject has recently been reviewed by Levin and by Whittaker and Feeny.

PLANTS AND INSECTS

Although the plant must utilize its chemical defenses against attackers and competitors of all sorts, here we shall examine only its relationship with insects. As insect repellents, the chemicals may operate in a number of different ways. Consider a hungry insect on a plant; it is there either because its mother laid it there as an egg or because it arrived there under its own power. The plant remains immobile, unable to escape or attack. Within its tissues is a complex array of chemicals ready to interact with the insect's sensory, digestive, or metabolic systems.

The first encounter between the insect and the plant (aside from visual and mechanical) is olfactory. Nearly every plant has an odor of some sort, and insects possess an extremely acute and discriminating olfactory sense. If the odor is repellent, the insect refuses to bite into the plant (or lay its eggs upon it if it is an ovipositing female) and sooner or later departs. If the odor is not repellent, the insect takes a bite. This action immediately brings it into contact with chemicals affecting its sense of taste. If the taste is deterrent, the sample bite does not lead to feeding. The plant is spared.

This sequence of events is essentially similar to those we ourselves experience every time we sit down to eat, although we never stop to analyze it. If food smells bad, we do not eat it; if food that looks and smells acceptable tastes bad, we refuse it.

On the other hand, a plant may not be at all unpalatable, but may be extremely toxic. An unfortunate insect that feasts upon it does not progress very far nor inflict much damage before it sickens and dies. Toxicity also extends to vertebrate herbivores. Some buttercups cause convulsions in livestock; larkspurs can kill range cattle; foxglove contains cardiac glycosides leading to convulsive heart attacks; a single leaf of oleander could kill a man. Equally protected are those plants which possess chemicals that are indigestible either because some of them inactivate the insect's digestive enzymes or because they are chemicals for which the insect has no appropriate digestive enzymes.

Other indirect mechanisms are exemplified by the action of tannins. Tannins render mature oak leaves useless as food to some moth larvae and to vertebrates by binding proteins into indigestible complexes. Still other chemicals act even more insidiously. As mentioned in an earlier chapter, many plants contain insect hormones or hormone mimics. These have the potential for disrupting the normal growth and development of feeders.

Many of the chemicals we are discussing are present in the plant in varying amounts at all times. Others are manufactured on the spot the minute the insect bites. Hydrocyanic acid and benzaldehyde (oil of bitter almonds) are examples. These two compounds characteristic of the rose family do not exist in the plants as such. They arise from the action of an enzyme (emulsion) on a glycoside (amygdalin). Each of these precursors is insulated from the other in separate compartments of the cells. When an animal bites a leaf, the compartments are broken, the two mix, and the toxic products are produced.

Another reaction that may also serve as a defense mechanism has recently been discovered in potato and tomato plants. When leaves of these plants are wounded by biting Colorado potato beetles (or by mechanically wounding the tissue), the tissue rapidly makes and accumulates a compound that powerfully inhibits digestive enzymes (proteinases). Within a few hours the inhibitor also accumulates in adjacent intact leaves. By the mere act of eating, the insect is causing that which it eats to become less digestible. In other instances, the plant's chemical machinery operates more slowly. When the plant is attacked, it synthesizes compounds that reduce its susceptibility to further attack.

If plants were totally successful, their chemical defenses impregnable, there would be no herbivores. As in all warfare, successful defense stimulates innovations in offense which in turn stimulate new defensive techniques, ad infinitum. Obviously, herbivores and parasites have been able to evolve countermeasures to the plant's defenses. To appreciate these, it will be helpful to think now about the insect/plant relation from the insect's point of view. Even casual observation reveals that each insect is characteristically associated with certain plants. No insect eats every green plant, and no green plant is attacked by the same group of insects. Carrot caterpillars eat carrots, the bean weevil eats beans, the cabbage looper eats cabbage, and so on; and the cabbage is eaten by the cabbage looper, the cabbage worm, the cabbage aphids, and many others.

Some insects are restricted to a single closely related group of plants as, for example, monarch butterfly caterpillars to milkweeds and silver-spotted fritillary butterfly larvae to violets. Others are restricted to a wider but still limited diet, as exemplified by cabbage caterpillars on all plants of the cabbage family and nasturtiums and the Colorado potato beetle on nearly all members of the potato family. And some insects have an enormous range of diet, a behavior best illustrated by plague locusts and by the gypsy moth caterpillar, which eats at least sixty-five different species of plants.

These dietary preferences raise the interesting question of what regulates the feeding preference of herbivorous insects. Although we have accumulated an astonishing collection of facts, we still do not really understand the mechanism of preference. From time immemorial, it was recognized that plant feeders had an unerring "botanical instinct." It was not until 1910 that the foundation was laid for an investigation of this "instinct." A Dutch botanist, E. Verschaffelt, working in the Amsterdam Botanical Garden, demonstrated that the plants eaten by caterpillars of the large white cabbage butterfly all contained mustard oils (thiocyanates derived from the breakdown of mustard glycosides by the enzyme emulsion). From that time on, the role of specific chemicals in regulating feeding preferences became the focus of extensive research. Nearly sixty years after Verschaffelt's report, another Dutch biologist, Louis Schoonhoven, demonstrated by electrophysiological methods that the caterpillars possess a taste receptor highly sensitive to the mustard glycosides.

In the 1940s, it was proposed that certain secondary plant substances, which were not necessarily of any nutritional value to the insect, acted as token stimuli signalling to the insect the presence of the proper food plants. At one point, a great controversy arose as to whether feeding preferences were regulated by token secondary plant substances or by primary nutritive compounds.

The present evidence strongly supports the hypothesis that individual insects are guided to their food by chemicals that afford sensory stimulation, that may in themselves be nutrients, but in the majority of cases are secondary plant substances acting merely as token stimuli. In other words, the choice by individuals is not nutritionally based; however, only those individuals survive and reproduce whose selection results in their having acquired a nutritionally satisfying diet. Thus, the fact that each insect is in a harmonious and optimal balance with its plant does not mean that it chose it by nutritional cues, but that the insect evolved simultaneously a digestive system to utilize the particular chemical complex of certain plants together with a sensory and central neural system enabling it to recognize those plants.

The plant has evolved chemicals that are designed to put off or kill the insect. Thus, many plants are not attacked by insects because of repellents or deterrents, and plants lacking these are susceptible to attack. At the same time that the plant is evolving these chemical defenses, species of insects are evolving counter-measures. Some insects evolve a "resistance" to some repellents and deterrents, and other insects evolve resistance to others. For example, the tomato hornworm is deterred from eating carrot by chemicals elaborated by the carrot, while the parsley caterpillar is deterred from eating tomato by chemicals elaborated by tomato. Obviously, some insects have evolved more widespread sensory tolerance than others.

Many insects have gone further; they have evolved not only a tolerance of specific chemical deterrents and repellents, but they have turned that tolerance into a positive attractance. Mustard oils and glycosides, which presumably evolved in plants of the cabbage family as defense mechanisms, have become attractants to cabbage caterpillars and aphids, so that the plant's defenses have been turned against itself.

In the case of toxic chemicals, different insects have evolved different detoxifying mechanisms enabling them to survive on special plants. For example, some insects can eat tobacco because they are able to detoxify nicotine; others eat pyrethrum because they alone can detoxify it. One species of woolly bear moth (*Seirarctia echo*) is able as a caterpillar to eat cycads—plants which are lethal to most other insects and are also carcinogens and mutagens for vertebrates—because it can detoxify the glycoside cycasin.

Again, as with deterrents and repellents, some insects have

turned the plant poisons to their own use. A striking example is that of several small beetles (*Chrysolina*) that can eat species of willows containing the poison hypericin. This compound is toxic to many insects and can cause blindness and starvation in vertebrates. The beetles not only are able to detoxify hypericin, but they have special taste receptors on their legs which detect the presence of hypericin and trigger feeding. The beetles incorporate the poisons directly into their own bodies so that they themselves become impalatable or toxic to their predators. The best known case is that of caterpillars of the monarch butterfly (*Danaus plexippus*), which accumulates cardiac glycosides from milkweed. Bluejays on first eating these caterpillars vomit and soon learn to ignore them. Monarch caterpillars induced to feed on plants lacking glycosides are quite palatable to jays.

This expanding knowledge of the chemical interactions constantly occurring between plants, their predators, and their parasites begins to reveal to us some of the factors that make particular crops susceptible to particular insects.

SHARING THE CROPS

Up to this point in considering ways to protect crops, we have assumed that if insects are present where there are plants, they will eat them. Operating on this premise, the strategy has been to reduce the population of insects. It is not, however, the presence of insects in the agricultural environment that is undesirable; the problem is created by the fact that these insects eat plants. If their feeding could be prevented, it would make little difference how many lived on the farm. This consideration suggests that agricultural control could be achieved if insects were simply kept away from crops. There are at least three ways in which this might be accomplished: decoy insects away from crops with competing lures, as has already been described; render the plants unacceptable; alter the environment of crops in such a way as to make the farm an unfit place for insect habitation. This last technique will be discussed in the following chapter.

The most successful of the several techniques is that of making plants unacceptable. One way of accomplishing this is to spray plants with chemicals that prevent feeding. The disadvantages of this approach are the same ones that attend the application of insecticides: costs of continuous application, residues, possible toxicity, and contamination of the environment.

A great deal of research is being conducted, especially in Hungary, on feeding deterrents (also called antifeedants). The search is for compounds that are nontoxic, do not persist in the environment, and can be absorbed by the plant in order to obviate the necessity of repeated applications; however, more research needs to be done, and much still has to be learned, before deterrents are available for everyday use. It is more clever to let the plant work for its own protection, as it did before the advent of agriculture. This involves the development of genetic varieties that are resistant to attack.

Although authorities differ on the precise definition of resistance, the essence of resistance from an agricultural point of view is that the ability of a plant to provide the maximum raw material (fruit, seeds, roots, leaves, fiber, lumber, etc.) for which it is cultivated is not diminished in the presence of insects (or parasites) that would have deleterious effects were the plant not resistant.

There are three components to resistance: *Tolerance* enables a plant to grow and produce well in spite of being infested by insects. It represents an excellent balance between the plant and its predators; both survive. *Antibiosis* is represented by the presence of materials injurious to insects or the absence of nutrients or vitamins essential to the insect. Insects feeding on plants with antibiotic resistance either die or produce fewer or no progeny, or grow so slowly that their damage is negligible. The third component of resistance is *non-preference*. Plants showing this component are unacceptable to egg-laying females and to feeding insects. Of the three components, tolerance is apt to be the least useful from a practical point of view. Non-preference and antibiosis not only permit acceptable yields from crops, but they also result indirectly in a decrease in insect population. Tolerant varieties yield well but allow the insect population to remain at a high level.

From the very beginning of agriculture, man has selected plants for special characteristics. This practice has not only been the essence of agriculture; it has in modern times been one of the factors ensuring the enormous success of agriculture by developing varieties with very high yields. The development of high-yielding varieties is a mixed blessing because of the dangers, discussed earlier with respect to maize, of dependence on a single variety which could be wiped out by a novel catastrophe. Furthermore, high-yield varieties demand optimum fertilizing and cultural conditions, and such conditions are difficult to meet in developing countries.

In any case, the conscious development of resistant varieties does not have such an ancient history. In America in 1785, a farm journal mentioned that there was a variety of wheat that was resistant to the newly-introduced Hessian fly, but subsequently all trace of this variety was lost. One of the earliest spectacularly successful applications of the principle of resistance was the use in 1870 of grape vines resistant to grape phylloxera.

In any given population of plants (and animals), there are always some individuals that survive disease, predation, or adverse environmental factors better than their neighbors. One way to establish resistant varieties of plants is to be alert for the individual that survives devastating attacks in the field and to breed from this individual. A striking example of this approach is provided by the history of tomatoes resistant to certain fungi. Scientists working in Peru noticed in a sugar cane field a tomato plant that was nearly immune. From this one resistant plant, known as the Trujillo tomato, were developed more than thirty varieties of commercially resistant tomatoes.

Not always, however, do resistant individuals that have been selected by nature possess desirable agronomic characteristics. Plant breeders have, therefore, attempted to speed up the process of selection and to control it in order to produce varieties that are acceptable as well as resistant. In the past, emphasis was on high yield and other agronomically desirable characteristics; and such plants with these characteristics tended to be very susceptible. Eventually, more attention was paid to resistance; but despite successes, described below, the exploration of resistance to insects has, comparatively speaking, been minimal, primarily for a lack of long-sustained studies and lack of funds.

Although California began in 1881 the first search for varieties of plants resistant to insects, the work ceased ten years later. Since 1916, the Kansas State Agricultural Experiment Station and the late R. H. Painter of Kansas State University pioneered and led in this field. However, during the first twenty years of study, the budget for the entomological phase of the research rarely exceeded $500 per year!

FLY AND APHID CONTROL THROUGH RESISTANCE

Three outstanding successes of insect control through resistance have been with the Hessian fly, the wheat-stem sawfly, and the spotted alfalfa aphid. As a matter of fact, the use of resistant varieties constitutes the only effective method for controlling these insects. Less than fifty years ago, the Hessian fly was a major pest of wheat, and losses were estimated in the millions of dollars. Today, losses are minor, and seldom, if ever, does this insect destroy an entire field.

The Hessian fly (*Cecidomyia destructor*) looks like a diminutive crane fly. The female lays her eggs in the stems of wheat, and the irritation caused by the young larvae between the leaf and stalk causes the stem to swell and the leaves to wither and die. After its arrival in America in 1776, it spread quickly westward. By 1804, it was alleged to have caused more than a million dollars worth of damage. By 1860, it had reached Iowa and Minnesota. The damage caused by it during the nineteenth century was variable and sporadic. It destroyed anywhere from 20 percent to 100 percent of a field. There were good years and bad years. The old reports provide an enlightening picture of the waxing and waning of pest populations and the spotty rather than pandemic geographical occurrence.

Beginning about 1840, farmers discovered that by delaying the sowing of wheat beyond the usual September planting (till the end of November), they could thwart the egg-laying of the fly and so reduce damage. In 1942, through the pioneering efforts of Reginald Painter in Kansas, a resistant variety of wheat, Pawnee, was released to farmers. Since then, twenty-two other resistant varieties have been released. By 1964, eight and one-half million acres were sown in resistant wheat. A conservative estimate of the saving that year was $119 million. In 1950 in Kansas, it was estimated that the saving in that year alone would pay for all the expenses of operating the Kansas State Agricultural Experiment Station for at least twenty years! One entomologist calculated that research on fly resistant varieties returned $28 for each tax dollar spent on research, including the salaries of plant pathologists, plant breeders, entomologists, and costs of maintenance. So successful have these varieties been that the fly has become rare in some wheat-growing areas. It is probable that its population levels could be reduced to a minimum throughout the wheat-growing areas of the United States if only resistant varieties of wheat were sown.

Another pest of economic primacy to wheat farmers is the

wheat-stem sawfly. The larvae of this small wasp-like insect live their entire lives within the wheat stem, boring up and down and eventually cutting it at the base. (Some control could be effected by rotating wheat with oats and barley, planting non-preferred hosts in infested fields, and employing other cultural techniques. Real control—as we shall see—was not achieved, however, until resistant wheat was developed.)

The story of the wheat-stem sawfly begins in Canada in the early 1920s. This sawfly had become a pest of major importance, and all methods of control had failed to produce satisfactory results. When hope was at its lowest ebb, an entomologist, C. W. Farstad, and an agronomist, A. W. Platt, working for the Canadian Department of Agriculture, observed that wheat with solid pith-filled stems was less severely damaged than wheat with hollow stems. Pith interfered with the normal movement of the larvae because they could not feed adequately in such cramped quarters. In Portugal, wheat with solid stems had been developed to resist damage by wind along the exposed Portuguese coast. Tests showed that this variety was especially resistant to attack by sawflies. By crossing this variety with a high-yield commercial variety, Farstad and Platt were able to produce eventually an agronomically acceptable strain that was ninety percent resistant to sawflies.

The amount of seed of this variety, aptly named Rescue, was very limited. In 1944, one bushel was given by Canada to the United States. In twenty-four months, the Montana Agricultural Experimental Station increased it to 60,000 bushels, during which time this bushel and its progeny had a remarkable odyssey. In the winter of 1944-45, the bushel was shipped to Arizona for planting. This planting yielded 35 bushels of seed by the spring of 1945. They were returned to Montana, planted and harvested in the fall of the same year, yielding 877 bushels of seed. These 877 bushels were sent to Arizona, where by the spring of 1946 they produced 3,870 bushels. These went back to Montana. At one point a railroad strike threatened to strand the seed, but railroad and labor officials agreed to allow that one item to be transported. Within one day of arrival at a number of dispersal points, the seed was in the ground, albeit one month behind the normal time for sowing. With the help of a providential rainfall, the seed sprouted and in the fall yielded 60,000 bushels. The battle was won.

In 1948, losses had declined by $4 million. By a conservative estimate, the bushel of seed given by the Canadians at a development cost of only a few thousand dollars resulted in a saving of $40 million. It would have taken American scientists, working independently, approximately ten years to develop their own resistant variety. By 1966, the North Dakota Agricultural Station and the United States Department of Agriculture had developed a strain resistant to both the sawfly and rust.

Another successful chapter in the development of resistant crops concerns the spotted alfalfa aphid, a plant louse first discovered in New Mexico in 1954, whence in two years it spread across thirty states, inflicting damage approximating $81 million. A crash program of research and development compressed ten years' work into three and produced a resistant alfalfa, Moapa. This was soon followed by five other varieties which by 1963 were growing on 2 million acres. While estimates of savings are always very rough and tend to be optimistic, there can be little doubt that the order of magnitude is millions at a cost on the order of $30,000.

The other major crop in the United States that has profited from the development of resistant varieties is corn (maize). Its most destructive pest is the European corn borer, introduced from Hungary or Italy around 1917. The estimated loss attributed to it in 1949 was $350 million. A great difficulty in developing resistant lines of maize arises from the highly competitive commercialism associated with the production and sale of hybrid corn seed, the high-yield seed adapted for various growing conditions. The pedigrees of these seeds are jealously guarded secrets, and such secrecy is detrimental to research. The future is nonetheless encouraging as one of the research workers, P. Lugenbill, in the field of plant resistance has remarked: "Resistance to the corn borer is of primary importance in marketing and selling hybrid seed corn, because seed companies have been forced to develop hybrids that are resistant to borer damage in order to survive in the highly competitive seed corn business."

The development of crop plants resistant to attack by insects is simply a matter of speeding up evolution, of artificial selection replacing natural selection. As pointed out earlier, in the co-evolution of plants and insects, crop plants are no longer able to evolve defenses against insects by natural selection because man is doing all the selection. Although he is speeding up selection, as measured by the evolutionary timetable, the development of successful resistant varieties still requires about ten years of work. This is too long for our impatient agro-economic system to wait. Only when all other measures seem to fail does one turn to plant breeding. There is no arguing the low cost of development and the high savings when success is achieved. Still, the time is long.

The successful development of resistant varieties begins by the chance discovery of a resistant individual plant or strain of plants, plants that have genes that make them tolerant, toxic, or unacceptable to a particular insect. Thus far, the knowledge that we have about the defense mechanisms developed by plants and, on the insects' side, the mechanisms underlying host preference and selection, has not been of use in selecting and developing resistant varieties because we do not yet fully understand these interactions.

The empirical method, the method of screening thousands of seedlings by using insects as test organisms, does not require inordinate amounts of time. However, if we knew more about the mechanisms of resistance, the genetics of resistance, and the insect/plant relationship, we might be able to reduce the development time from ten years to a fraction of that. At the moment, however, the amount of financial support allocated to this approach is minuscle.

SHORTCOMINGS OF RESISTANCE

We cannot leave the subject of resistance without mentioning some of its shortcomings. One of these is the difficulty of producing varieties that are resistant to all pests. Basic studies have suggested why this is so difficult. A particular plant evolves a complex array of defense mechanisms: Each is effective against different enemies. The plant geneticist, therefore, is faced with the problem of breeding for multiple characters. There is also the danger that in selecting a character that provides resistance against one pest, the plant breeder may be making the plant more susceptible to other pests. A hint of this potential danger has already come from studies of resistance in cucumbers.

Cucumbers normally produce a class of terpenoid compounds

called cucurbitacins. These compounds are specific feeding attractants for cucumber beetles of the family Chrysomelidae. There is a direct quantitative relationship between cucurbitacins and damage inflicted by the beetles; therefore, varieties lacking the compounds would not suffer attack. Cucumbers, however, have other predators that are non-specific. Among these are two-spotted mites (*Tetranychus urticae*). When normal bitter cucumbers and artificially bred non-bitter cucumbers were experimentally infested with mites, it was discovered that the non-bitter plants later had more mites and suffered greater damage. The bitter plants were relatively free. The bitter plants deterred feeding, delayed it, and increased mortality.

Here we have an excellent demonstration of a plant having evolved secondary substances that protect it against predators (mites in this example), and an insect, the cucumber beetle, that has evolved the means of using the compound as an attractant. The beetle thus insures itself a feeding advantage because competition is removed. This case also demonstrates that a plant must evolve multiple mechanisms for protection against different enemies. From the practical point of view of resistance, it demonstrates the difficulty of developing varieties resistant to one pest without at the same time making the plant more vulnerable to others.

There is one more point that must be considered. Just as insects can become resistant to insecticides and to parasites in biological control, they can become resistant to resistant plants. The mill of evolution grinds on. The plant breeder "evolves" new strains of plants; the insect responds by evolving strains that can attack it. This is happening, for example, with the Hessian fly. For each dominant resistant gene in wheat there is a complementary recessive gene for survival in the fly.

In Indiana, four different races of flies have been found in natural field populations that are able to infest and reproduce successfully on varieties of wheat with different types of resistance to Hessian fly. Originally race A predominated in the fields of Indiana. When strains of wheat resistant to this race were planted, the population of that race declined, and race B, which could attack the resistant wheat, increased. This development had been anticipated and farmers were given a new variety of wheat that was resistant to race B. New biotypes also developed in the green bug (*Schizaphis graminum*), a pest of sorghum, in the corn leaf aphid (*Rhopalosiphum maidis*), and in the spotted alfalfa aphid. This is reminiscent of human experience with flu vaccines. No sooner had medicine provided us with vaccine for one type of flu than another virulent type appeared. Fortunately, this is not a universal phenomenon. Some varieties of maize resistant to European corn borer (*Ostrinia nubilalis*) that have been grown continuously since 1949 are still resistant. There is also a variety of apple, Winter Majetin, that was resistant to woolly apple aphids in 1831, and still is.

Developers of resistant varieties of plants cannot rest on their laurels. Ideally they should develop varieties resistant to all known strains of an insect. Even though this development is extremely difficult, some multirace-resistant varieties have already been prepared for release to wheat farmers. One optimistic note emerges from the consideration of the time scale of the various processes involved. Even though it takes on the average ten years to develop a resistant variety of crop, the process has speeded up evolution a hundred or more times. The insect may not be able to evolve that rapidly in response. If this is so, the scientist can establish a comfortable headstart. By constant vigilance and research he ought to be able to maintain this advantage once established.

The employment of resistant varieties for insect control has many advantages, not the least of which is economics. Figures quoted by P. Lugenbill show impressive returns. Costs of development of resistant wheats come to 115 professional man-years for the Hessian fly; 92 for the stem sawfly; 119 for the alfalfa aphid; and 136 for the corn borer. At $20,000 per man-year, this totalled $9.3 million invested by federal, state, and private agencies. The farmer saved $308 a year. A resistant variety is effective for about ten years. Then it is discarded and replaced for biological, agronomic, or other reasons. Over ten years the accumulated net saving is about $3 billion, a return of 300:1 on the research dollar invested. This value does not include such bonuses as the eradication or suppression of the local insect population or the numerous savings accruing from the elimination of spraying and of residues. As Lugenbill concludes: "Resistance to pests exists throughout nature. It is all around us in animals and plants. It needs only to be discovered and put to work to solve many of our most serious pest problems."

References

1. Beroza, M. (ed.), *Chemicals Controlling Insect Behavior*. Academic Press, New York and London, 1970, 170 pp.
2. Da Costa, C., and Jones, C. M., Cucumber beetle resistance and mite susceptibility controlled by the bitter gene in *Cucumis sativus* L. Science, 172 (1971), 1145-46.
3. Dethier, V. G., *Chemical Insect Attractants and Repellents*. Blakiston, Philadelphia, 1947, 289 pp.
4. Dethier, V. G., Host plant perception in phytophagous insects. Symp. Intern Congr. Entom. 9th, Amsterdam, 1951, 2 (1953), 81-88.
5. Dethier, V. G., Evolution of feeding preference in phytophagous insects. Evolution, 8 (1954), 33-54.
6. Dethier, V. G., Feeding behaviour. In: *Insect Behaviour* (P. T. Haskell, ed.), Roy. Entom. Soc., London, 1966, pp. 46-58.
7. Dethier, V. G., Chemical interactions between plants and insects. In: *Chemical Ecology* (E. Sondheimer, and J. B. Simeone, eds), Academic Press, New York, 1970, pp. 83-102.
8. Ehrlich, P. R., and Raven, P. H., Butterflies and plants; a study in coevolution. Evolution, 18 (1965), 586-608.
9. Fraenkel, G. S., The *raison d'être* of secondary plant substances. Science, 129 (1959), 1466-70.
10. Green, T. R., and Ryan, C. A., Wound-induced proteinase inhibitor in plant leaves: a possible defense mechanism against insects. Science, 175 (1972), 776-77.
11. Hatchett, J. H., and Gallun, R. L., Genetics of the ability of the Hessian fly to survive on wheats having different genes for resistance. Ann. Ent. Soc. Amer., 65 (1970), 1400-07.
12. Jermy, T., Biological background and outlook of the antifeedant approach to insect control. Acta Phytopath. Acad. Sci. Hungaricae, 6 (1971), 253-60.
13. Jermy, T., and Matolesy, G., Antifeeding effect of some systemic compounds on chewing phytophagous insects. Acta Phytopath. Acad. Sci. Hungaricae, 2 (1967), 219-24.
14. Levin, D. A., Plant phenolics: an ecological perspective. Am. Naturalist, 105 (1971), 157-81.
15. Lugenbill, P., Developing resistant plants—the ideal method of controlling insects. U.S.D.A., Agri. Res. Serv. Prod. Res. Rpt. 111, 1969, pp. 1-14.

16. Merz, E., Pflanzen und Raupen: Über einige Prinzipien der Futterwahl bei Gross-schmetterlingsraupen. Biol. Zbl., 78 (1959), 152-88.
17. Painter, R. H., Insect Resistance. In: *Crop Plants*, MacMillan, New York, 1951, 520 pp.
18. Pathak, M. D., Genetics of plants in pest management. In: *Concepts of Pest Management* (R. L. Rabb and F. E. Guthrie, eds.). North Carolina State University Press, Raleigh, 1970, pp. 138-57.
19. Riley, C. V., Third Rep. U. S. Ent. Commission. Washington, D.C., 1883, 347 pp.
20. Schoonhoven, L. M., Chemosensory basis of host plant selection. Ann. Rev. Ent., 13 (1968), 115-36.
21. Thorsteinson, A. J., Host selection in phytophagous insects. Ann. Rev. Ent., 5 (1960), 193-218.
22. Whittaker, R. H., and Feeny, P. P., Allelochemics: chemical interactions between species. Science, 171 (1971), 757-70.

Chapter Eight
Toward Wisdom

8
Toward Wisdom

The living world is a dynamic system. The myriad forms of life that coexist on the planet struggle for a share of its limited resources. All of the species that surround us have come to terms with their environment. Those that have not are extinct.

When man began to shape the environment to his own design, introducing changes wittingly and unwittingly, the impact was felt by all other organisms. Because the biosphere is dynamic and because organisms are responsive to changes therein, they began to adapt to new conditions and even to exploit them. Changes in the relation of insects to plants, changes in the levels of population of insects, changes in their distribution, their behavior and their genetics are normal inherent expressions of living things. Students of evolution are not at all surprised at the results of man's activities. Only those who have not looked insightfully into the essence of living have been surprised.

Insects will respond to change for as long as they exist. Their populations will fluctuate and will adapt, and one face of adaptation is resistance. It would be a mistake to think that there will be resistance only to insecticides. There are already examples of resistance to chemosterilants, hormones, resistant varieties of plants, and to introduced parasites. How then can man successfully maintain the artificial imbalance in nature that is agriculture?

There is no turning back the pages of history. The face of the earth cannot be returned to its pre-human visage. Once introduced, insects and plants can not be extradited to their ancient homelands. Agriculture can not be returned to its primitive hoe culture, when plants were grown for food and not for profit. We have to live with other species in the world we have created. We have to abandon, or, hopefully, reverse, abiotic conditions of our own making that are strangling us; but we have to coexist on compromise terms with other forms of life. Specifically, we have to eliminate all forms of pollution on the one hand while preserving a favorable population balance with plants and animals, especially insects, on the other. Eradication of anything but ourselves is a myth. Our dilemma then is how to manage plants and insects without annihilating the environment.

NEW SOLUTIONS TO OLD PROBLEMS

We have tried many approaches and failed. Insecticides have failed not because of any inherent weakness in the concept of reducing insect populations by chemicals. They have failed because of misuse, because of the unrealistic goals we set ourselves, because of irresponsibility, profit motive, laziness, and ignorance. Biological control has been unsatisfactory mostly because of the feebleness of efforts in research and development, lack of financing, and the obsession with chemical control. The use of resistant varieties of crops has enjoyed only limited success for similar reasons. The more exotic methods of control, hormones, pheromones, genetic manipulation, and others are, with one or two exceptions, still in the early experimental stage.

One method that attempts to compensate for the artificiality of modern agriculture is basically ecological. From earliest times, farmers employed ecological principles to control crop pests even though they may not have understood the meaning of "ecology." In colonial America, for example, farmers were advised to burn stubble and crop debris, to rotate crops, to adjust times of planting, to practice clean cultivation, and to manure heavily. Many of these practices still make sense. The insecticides available at the time were not so effective as to discourage the farmer from continuing sensible cultivation practices. After World War II, however, many sound ecological practices were abandoned because DDT and other powerful insecticides could be applied less laboriously and appeared to be more efficient. Their power was so great that they could counteract, for awhile, the effects of sloppy agriculture. Only when the revolution against insecticides began was there any sentiment for returning to more rational ecological practices.

Of the several artificial conditions of agriculture that help create a pest problem, monoculturing, the establishment of huge expanses of one kind of plant in a sterile, weedless expanse of soil, is one. In addition to their shortcomings from an ecological point of view, these monocultures lack the age distribution of natural populations (all plants are of the same age), and they are not self-perpetuating; life begins when the farmer sows the seed and ends when he harvests. One way to restore stability to these artificial ecosystems is to reinstate some of the conditions of natural ecosystems. Diversity is glaringly absent in monocultures. What this means is that there are no alternate host plants for pest insects; nor are there salubrious environments for parasites and predators.

Diversity can be established by intermingling crop plants with other plants; however, diversity is a double-edged sword. While it can benefit the farmer in some cases, it can defeat him in others. Two of the most critical factors which determine which way the sword cuts are the seasonal characteristics of a crop (that is, whether it is an annual or a perennial) and the climate of the area.

Perennial crops tend to create more stable conditions because their pests are less prone to disperse. The pests, their enemies, and the crop tend to form a more or less permanent closed system. The task of the agriculturalist is to stabilize the situation still further in a direction favorable to the grower. Orchards fall into the category of permanent closed systems. By contrast, annual crops are usually populated by pests that immigrate when the crop appears. No population of predators and parasites immigrates at the same time; therefore, the number of pests builds up to damaging proportions before any natural checks intervene.

Some stability can be obtained in this case by artificially introducing parasites *and pests*. This has been done with some success with strawberries and with certain cruciferous crops (e.g., cabbage). The point is that the introduction of *some* pests along with the parasites provides breeding ground for the parasites, which can then multiply to cope with pests that will immigrate naturally.

Diversity can be introduced by intermingling other plants. Control of the cotton boll worm has been achieved in Peru by interplanting maize and cotton in irrigated areas. The climate is equable. Here the boll worm has several generations, and while it is growing on maize, there is time for its enemies to build up to high levels. A similar situation prevails in Uganda, where the climate is equable also and where the boll worm breeds (with its enemies) throughout the year on wild plants adjacent to the cotton. In Tanzania, however, there is a dry season, during which time the enemies hibernate (diapause), so that planting maize with cotton merely provides more food for the boll worm and causes large numbers to grow on maize after which they migrate to the cotton.

The influence of climatic conditions is also illustrated by comparing the growing of cruciferous (cabbage, broccoli, etc.) crops in South Africa with their culture in Britain. In South Africa, where the climate is equable, it pays to grow cruciferous crops in rapid succession because this ensures a constant population of enemies. In Britain, where there is a winter break in the development of insects, it is a mistake to keep cultivation going, especially between September and April, because all the overwintering parasites and predators are destroyed by disturbing the soil.

In some situations, diversity helps, in others it hinders. The black bean aphis (*Aphis fabae*) spends part of its life on beans and part on other plants. If these wild alternate hosts were removed from hedgerows and other uncultivated areas adjacent to beans, the pest population would be markedly reduced or perhaps locally eliminated. On the other hand, removal of flora from hedgerows also depletes predator populations.

In California vineyards, control of a leaf-hopper was achieved by adding small areas of a special blackberry. In Tanzania, control of one of the coconut pests was achieved by intermingling other species of plants with the coconut palms. These provide nesting places for an important predatory ant and a source of food for coccids, whose honeydew is in turn an additional source of food for the ant. The mixed coconut grove then becomes a congenial home for an effective predator.

Controlling diversity is only one of many cultural practices that can reduce the severity of insect infestation. Spacing of crops, destruction of volunteer plants, replacement of alternate favored hosts, strip harvesting, timing of harvesting, timing of planting, regulation of irrigation, modification of fertilizing, and agricultural sanitation are a few of the many commonsense ways of compensating for the artificiality of agriculture or of capitalizing on weak links in the life histories of insects. Some of the weak links are: a necessity to hibernate, a need for specific day lengths, a need to alternate hosts, a sensitivity to critical temperature limits, and a preference for particular microhabitats.

Agricultural sanitation is a powerful aid in the control of many insects. Since 1897, control of European corn borer in Europe was assisted by the destruction of the crop residue in which the borers spent the winter. In the United States, many farmers husk, shred, ensile, or plow under the remains of the crop after harvesting. In the case of the pink bollworm of cotton, community-wide stalk-shredding followed by plowing under can reduce populations more than 95 percent. Tillage carried out properly at the right time of year is another sensible agricultural practice that can raise havoc with overwintering populations of many crop pests. In North Dakota, for example, populations of the wheat-stem sawfly have been reduced 75 percent by tillage. Many grass-hopper eggs are also destroyed. These and other cultural practices are frequently closely associated with ordinary farm practice, hence are simple and cheap and require no extra outlay of equipment. Many are matters of common sense; others depend on at least a cursory knowledge of the life histories and behavior of the insects involved.

We could go on, but the point should be clear. By removing some of the artificiality from agricultural systems, it is possible to reduce many pests to a tolerable level. This means that there must be acceptable tolerable levels in the first place. The idea of eradication must be abandoned; the grower must adopt a psychology of prevention rather than cure. He must accept the idea of prevention long before infestation actually occurs, and he must time his operation carefully.

Cultural practices, however, seldom generate *spectacular* results; therefore, they lack the appeal of chemical treatment. The highly specialized mechanization introduced in modern agriculture to sow, cultivate, and pick crops is also a source of difficulty. Many of the techniques work only under the particular conditions for which they were designed. Interplanting, for example, could interfere with some to the point where more extensive methods of harvesting would be required. The farmer compares the costs of the two methods, but he seldom allows for the cost of control that must accompany the intrinsically cheaper method. Nor does he look at the cost over a span of time. It is argued that cultural practices must be compatible with sound agronomic practices. Yet, the soundness of an agronomic practice must be weighed against the soundness of a cultural control practice. A trade-off may be necessary. Then there is a hard choice to make.

The hard choice always emerges. On-the-spot profit always

looks more attractive than long-range benefits. For example, wheat bulb fly (*Leptohylemyia coarctata*) can be controlled by sowing early, preferably on land that has not fallowed; however, farmers will grow wheat after a crop of potatoes because they get $40-$50 per hectare more for wheat than for barley, even with the rising costs of the insecticidal treatment necessary to produce wheat under these conditions. Until very recently, ecological manipulation had very little impact because it involved labor practices that were unprofitable or ill-suited to the highly specialized technology of mass farming.

Ecological control is especially relevant to forest insects. Serious outbreaks are a natural feature of large coniferous forests. They may represent a natural way of keeping a forest in balance and ensuring its healthy continuity because the insect populations build up after natural disasters to the forest—fire, wind, and drought—and also when trees become diseased, crowded, and aged. Timber interests, however, see the destruction of trees as a loss in revenue, and recreational interests see the destruction as detrimental to tourist traffic. Also, these seems to be a sentiment against allowing the forest to go through its long, slow, and sometimes unsightly cycle.

In 1971, a National Park Service biologist voiced the new philosophy: "Attempts at 'control' are not compatible with the objectives of natural areas, and the deaths of susceptible trees by native insects should be recognized as a natural process. The mountain pine beetle. . .is responding to favorable environmental conditions in portions of Yellowstone National Park and is attacking stands of lodgepole pine (*Pinus contorta*). No attempts at control are being considered. . . ."

Not too long ago, the attempts to halt these cycles were brutal. During 1924-27, an area ten miles long by 150 yards wide in Yellowstone Park was infested with lodgepole sawfly and pine tube moth. Each year for four years the area was doused with 3,800 pounds of arsenate of lead mixed with 140 gallons of fish oil and 60,800 gallons of water. This amounted to 2.23 tons per square mile, leaving, over four years, a residue in the soil of 27.88 pounds per acre of insoluble lead arsenate!

Today, silvicultural methods and more reasonable attitudes prevail in most instances. Recognizing, for example, that mountain pine beetles (*Dendroctonus ponderosae*) wreak their worst damage in overstocked second growth, the Forest Service recommends precommercial and commercial thinning as a preventive measure in ponderosa stands, and felling, piling, and burning plus spraying as a control measure. In lodgepole stands, it recommends logging as the only practical method of control, and the maintenance of robust trees by shortening the rotation of crops.

As is the case with food crops, each species of tree and each pest requires its own prescription. In many instances, prevention and control are achieved through cleaner cultural methods, clear-cutting infested stands, cleaning up, and salvaging. Sprays and parasites are still recommended in some instances. Spraying should be the last resort, not the first.

The present situation with regard to the management of forests is described with wry humor by J. H. Borden:

If I were asked to classify present-day forest-insect managers into general categories, I would have to say that the intelligent majority are scared stiff or confused (or both)! All the rest are still living in the first half of the century and greatly oversimplify the problems which they attempt to solve. They are not aware of or they ignore the volumes of work which have disclosed the tremendous complexity of living systems and the subtle forces which ensure a balance of nature. Moreover, they are ignorant of or unconcerned about the consequences of any pest-management action they may take.

Why are the more intelligent forest-insect managers scared stiff and/or confused? Simply because they have enough information at their disposal that they now recognize the complexity of the problems they face. They realize that they neither completely understand the ecosystems in which they work nor can they accurately predict the consequences of their pest management endeavors. The situation in which forest-insect managers increasingly find themselves was honestly and clearly voiced by one beleaguered forest entomologist. . .when he said "It was apparent that I am in a dilemma about 93 percent of the time, and the other 7 percent of the time I am on annual leave and thinking of something else."

None of the methods of control available to us are intrinsically ineffective. It is only our employment of them that has been ineffective. One basic flaw has been the "either/or" attitude. Different techniques have in the past been championed by different people. Methods have been viewed as alternatives to one another. The enormous biological complexity of the web of life together with the equally staggering ecological, economic, and sociological interactions of agriculture indicate very clearly that there is no one method applicable to all situations nor even one method applicable to any particular system. Part of the solution lies in integrating all systems—chemical, biological, and ecological—into one approach and in an association that is sufficiently flexible to permit modification. There can be no single integrated approach. Approaches must be tailor-made to fit each situation.

TOWARD INTEGRATED CONTROL

The inspiration for integrated control arose simultaneously about twenty-five years ago in Germany, Nova Scotia, and California. In their forests, the Germans conducted a sophisticated program of supervised control in which no control operations were begun until the role of natural mortality factors had been painstakingly evaluated. In Nova Scotia, A. D. Pickett and his co-workers pioneered the ecological approach to control of orchard pests. Their credo was that a thorough knowledge of population dynamics and ecology would suggest a program of control that would work with insecticides *and* natural enemies. During one period, 80 percent of the acreage devoted to apples and pears was controlled this way, successfully and with spectacular savings in expense. (A sad sequel is that increased demand for blemish-free fruit and changes in economics have caused the growers to return to straight chemical control.)

California, the very place described a few generations ago by the entomologist L. O. Howard as the hotbed of crackpots and crackpot ideas, nurtured and took the lead in integrated control as it already had in biological control. The late Professor H. S. Smith may be considered the pioneer because of the profound influence on his students and later entomologists through his philosophy of viewing pest populations as ecological phenomena. R. F. Smith and R. van den Bosch crystallized and preached the idea of integrating all techniques for control. In Canada, the concept also evolved from the work of B. P. Beirne and his colleagues. Clark, Geier, and Hughes initiated and championed integrated control in Australia.

Integrated control initially meant the combined use of insecticides and natural enemies in a balanced program. Gradually the concept was expanded to include, in addition, ecological and cultural control. Its philosophy is to manage a pest population rather than eradicate it and to manipulate by integrated techniques as many of the variables as possible that influence the economic injury level caused by pests. In this connection, insecticides should be fitted into the ecosystem, not imposed upon it. Despite all their drawbacks, chemicals still remain a powerful, economical, strategic tool if used effectively, wisely, and as an adjunct to other measures. Banning *all* insecticides is not intelligent.

The application of integrated control is incompatible with the carefully nurtured public idea of a pest-free environment. The two philosophies are diametrically opposed. Not only the public, but also a very large segment of the entomological profession, and nearly all of the pesticide manufacturers, fiercely argue that our only hope for agricultural survival lies in eradication, or at least reduction of insect populations to *minimal* levels. Proponents of integrated control argue that we should *manage*, not eradicate, populations, that we should accept tolerable levels. A new term descriptive of control has been proposed in the past few years, a term which at least semantically permits the two opposing philosophies to coexist. This is Pest Management. Pest Management encompasses all of the concepts of integrated control but does not exclude the possibility of eradication.

The concept of pest management did not begin to receive financial support until 1972. Funds were supplied by the United States International Biological Program through the National Science Foundation and by the United States Department of Agriculture. One pilot program for alfalfa pest control is already in its third year of development at Purdue University in Indiana (5). It is an excellent example of how the best biological, cultural, and chemical technologies can be integrated, how a continuous flow of information about regional insect population patterns and environmental factors is employed, and how predictions can be made by combining daily weather data, field sampling, and advanced computer technology which can simulate actual conditions.

The road ahead for pest management (or integrated control) is difficult because it attempts to deal in a sophisticated way with very complex ecological relations, because it requires the cooperation of experts from many different fields of study, because there is a shortage of trained professionals to advise growers, and because there is still a dearth of basic information. Nevertheless, it offers the best hope for protecting crops and preserving the environment.

In any case, we have to revise our goals and our philosophies. First we must learn to live with insects. No animal can live a robust healthy life in a germ-free environment, as has been demonstrated by raising mice under such conditions. No more can we live in an insect-free environment. We must recant the belief that "the only good bug is a dead bug." We must reject the unnatural notion that all of our food must be cosmetically perfect. This insidious creed fostered and perpetuated by governmental agencies, food-processing industries, and purveyors of produce has done incalculable harm. It is akin to obsessive hand-washing, a clear psychopathology.

We must accept the idea that agriculture is now "big business" governed by profit motives even in developing countries, and that malnutrition and famine are not the immediate forces that govern agricultural practices. We must reconcile the socioeconomic factors with the necessity of introducing a time factor, that is to say, we must amortize capital, investment, pest control procedures, and consumer demands.

For example, there are times when it may be better to bear a temporary loss rather than impose heroic control measures (e.g., aerial spraying of thousands of acres of forest) that give an immediate saving and a long-range loss. It is unrealistic to expect the individual farmer to suffer personal loss for the sake of society unless society is willing to make comparable sacrifices for the farmer. Some consideration might be given to subsidizing losses incurred by *not spraying* a crop (in the same way that price support is applied) in the interest of obtaining long-range ecological and environmental stabilization. It might even turn out that this step would be less expensive than "insurance" spraying.

We should consider removing the sale and dispersal of insecticides from the control of industry. This change may provide the only way to prevent abuses. Above all, control must be taken out of the hands of the amateurs who now are in power and placed in the hands of professionals. It is frightening to learn how powerful uninformed people and people with vested interests can become. Even members of vacation bureaus are assessing situations and trying to make policies. On November 11, 1973, the *New York Times* reported, for example, as follows:

The Pocono Mountains Vacation Bureau, representing scores of resorts in this region, has started a campaign to appeal to Congress to have the Federal ban on DDT lifted temporarily to permit the spraying of area forests.

The agency's president, Paul Asure, contends that "DDT is the only hope left" in combating the gypsy moth, which is devastating the forests of four Pocono Mountain counties.

"Thousands of acres of our beautiful Pocono forests have been defoliated over the past few years," Mr. Asure said in announcing the campaign to have DDT used in the resort area next spring. He said that further destruction of the forests could threaten the future of the area as a resort center.

Professionals must, however, be more widely trained than is presently the case. In addition to having sound biological training and a basis in integrated control, they must be as fully aware of the social and economic sides of the problem as of its ecological side. Control must be based on sound biological principles, *proven* biological techniques, but must be governed by economic principles. It must eschew politics, vested interests, emotion, and propaganda.

Admittedly, we do not live in a Utopia in which all these conditions are met. For example, there are not enough professionally trained people to man such programs. R. L. Rabb (5), in introducing a conference on pest control, summed up this aspect of the situation as follows:

The status seems to be that despite a considerable literature bearing on pest management there are relatively few individuals with adequate knowledge and experience active in the field. Many factors are responsible for this unfortunate situation. The public image of the applied entomologist has suffered during the past decade or two by unfavorable associations with rather unscientific *ad hoc* methods of control and environmental pollution. Our young people have not been exposed to the true nature of the field, which is just as exciting, intellectually challenging, and urgently relevant to mankind as any discipline. Administrators must also share the responsibility for the slow pace with which traditional organization of research is being modified to accommodate the new type of research projects conducive to the development of pest management.

An integrated team approach oriented to a pest complex, rather than a mosaic of small individualistic one-commodity projects, is needed to develop the expertise for managing natural populations of insects. The talents of the scientific community presently being brought to bear on pest control is pitifully inadequate, and consequently, some very important questions must be answered before significant progress can be expected.

Furthermore, we face the problem of convincing the grower that an integrated program is to his best interest. He wants to maximize his profits with the least expenditure of energy. At the same time he is under the spell of industry. Ray Smith summed it up this way (ɔ):

Sometimes I become very discouraged about this aspect of the development of integrated control. Not too long ago someone said that changes do not come about unless there is a crisis; hence, until the farmer is in a very desperate situation he is not looking for alternatives. Integrated control is not a simple easier system; integrated control is a more difficult and sophisticated system. It doesn't come about by taking the easiest way. Also, although I have a lot of friends and I have a lot of respect for many people in the chemical pesticide industry, there are certain parts of that industry which stand in the way of the development of better pest control. This is mainly at the level of the local pesticide distributors and sellers and not at the higher echelons of industry; but nevertheless, the block exists. It remains as a problem for all agriculture which stands in the way of the development of better pest management.

Again the economic constraint comes to the fore in several of its many guises. There is the economic pressure from the chemical industries and their associates and the economic motives of the farmer. With regard to the feasibiltiy of integrated control, some authorities as, for example, P. S. Corbet of Canada, feel that integrated control can not compete economically with other methods and that its lack of economic competitiveness can not be removed until agriculture and economics get uncoupled, in other words, as long as the goal of agriculture is to maintain a high sustained yield with the least expenditure of energy.

As it is, the modern farm is an energy sink. An estimate of the amount of fuel used in Great Britain to power tractors for sowing, tilling, and harvesting and to produce fertilizers and pesticides reveals that British agriculture is using three times as much fossil energy as the energy in harvested crops. The modern American farm uses five or six times more fuel calories than are harvested as food calories. According to various estimates, some as low as 35 years, the fossil fuel age of agriculture has a limited future.

Since the methods available to us for controlling agricultural pests involve so many factors over and above agricultural ones, we have to ask ourselves several hard questions before deciding upon any strategy. What standard of living do we wish to attain? What price—monetary, social, health-wise, and environmental— are we willing to pay? What are our priorities? What is our time scale? In opting for solutions, are we thinking in terms of our generation or of generations yet unborn?

In the end, cliché or not, the solution lies with an educated and concerned public. Probably the most fundamental difference between the insect and ourselves is that the insect, with little capacity for learning, survives by his extraordinary adaptability to all environments and all changes that man engineers, while man survives because of his superior learning ability. Clearly we have the intellectual ability to triumph over the insects' superior adaptive ability. We may also already possess the knowledge to answer the hard questions and solve some of the pressing problems. But do we have the wisdom and, above all, the unselfishness to use this knowledge effectively for the benefit of mankind?

References

1. Anon., Principles of Plant and Animal Pest Control. Vol. 3. Insect-pest management and control. National Academy of Sciences. Washington, D.C., 1969, 508 pp.
2. Borden, J. H., Changing philosophy in forest-insect management. Bull. Ent. Soc. Amer., (1972), 268-73.
3. Davidson, A. G., and Prentice, R. M. (eds.), Important forest insects and diseases of mutual concern to Canada, the United States and Mexico. Roger Duhamel, Ottawa, 1967, 248 pp.
4. Dolph, R. E., Mountain pine beetle damage in the Pacific northwest 1955-1966. Forest Service, U.S.D.A. (no date).
5. Rabb, R. L., and Guthrie, F. E., *Concepts of Pest Management*. North Carolina State University, Raleigh, N. C., 1970, 242 pp.
6. Stern, V. M., Smith, R. F., Van den Bosch, R., and Hagen, K. S., The integrated control concept. Hilgardia, 29 (1959), 81-101.

Index